T0245191

Food product design

Food product design

An integrated approach

edited by:
Anita R. Linnemann
Catharina G.P.H. Schroën
Martinus A.J.S. van Boekel

Wageningen Academic
P u b l i s h e r s

This work is subject to copyright. All rights are reserved, whether the whole or part of the material is concerned. Nothing from this publication may be translated, reproduced, stored in a computerised system or published in any form or in any manner, including electronic, mechanical, reprographic or photographic, without prior written permission from the publisher:
Wageningen Academic Publishers
P.O. Box 220
6700 AE Wageningen
The Netherlands
www.WageningenAcademic.com
copyright@WageningenAcademic.com

The individual contributions in this publication and any liabilities arising from them remain the responsibility of the authors.

The publisher is not responsible for possible damages, which could be a result of content derived from this publication.

ISBN: 978-90-8686-173-6

First published, 2007
Second, revised edition, 2011

© Wageningen Academic Publishers
The Netherlands, 2011

Table of contents

Foreword

Historically, the food industry centred on the availability of simple raw materials and a limited range of appropriate processing technologies. This has resulted in a supply-driven food manufacturing industry, providing the consumer with a limited number of options and product choices. In more recent times, supermarkets have become the main power-players in the global food industry, because of their controlling linkage between food manufacturers and the consumer. Consumers have dramatically increased their awareness of the global variety of foods and the link between health and the food we eat. There has been a radical evolution of social responsibility (environment, etc.) and realisation by consumers that they can actually demand foods in formats and presentations they had never seen before. Concomitantly, the emergence of dramatic new technologies (genetic engineering, IT, processing, packaging) and innovative industrial "intermediate" ingredients has completely changed the manufacturing environment and the way the food industry confronts innovation.

Today, the consumer is the starting point for food product and process development. Food technologists don't have time to conduct statistically-explicit experimental product design procedures that were a feature of 20th Century product development. Efficient and effective shortcuts are essential to control costs, achieve suitable speed to market, and to assure consistent, shelf stable, safe and nutritious food products.

A modern supermarket contains over 20,000 stock keeping units of food products, of which 90% did not exist 10 years ago. Out of the tens of thousands of new food products that are offered each year, more than 5,000 new products are introduced to these shelves, but more than 70% are discontinued within 3 years. The life cycle of many modern food products is around 6 months.

The modern supermarket is driven solely by turnover of products from its shelves. Given the supreme power of the supermarket today and the direct relationship between consumer preferences and supermarket turnover, it is logical to conclude that the 21st Century consumer is king (or queen). This MUST be recognised as the starting point of any modern-day food product and process development practice.

Because of the enormous cost of introducing new food products to the market and the economic value of product failure it is essential that modern businesses use all possible commercial acumen to make decisions about product development, and adequately support it. A company's ability to remove unnecessary costs from the conversion of raw materials to consumer products requires special technical and technological abilities, which are much more complex than even 10 years ago. The industry is evolving rapidly and novelty along the entire food chain – from the soil to the consumer – occurs at an incredible rate. Smart product development is essential to remain in business. While failure to innovate is a sure recipe for company failure, it is equally true that failure to innovate in a smart way will add commercially unviable costs to a company, and the result will be as devastating as not innovating at all.

This book assumes a sound knowledge of food science and engineering. Its focus is on food technology – the application of science and engineering to a commercial industrial setting and it demonstrates the integration of existing knowledge and its application in a pragmatic and effective manner. Modern tools, available through the evolution of information technology, are an extremely effective way to innovate the product development process itself. The application of such techniques is explained clearly, concisely and effectively both for those working towards entering the industry, and for those already active.

Wageningen University in the Netherlands, like Massey University in New Zealand, is a bastion of food technology education and training for its hemisphere. Its focus is unashamedly applied, pragmatic and aimed at the food manufacturing industry. At Wageningen University the understanding of fundamental science and engineering is leveraged for application in a commercial environment. The University's industrial linkages are paramount in both the research focus and the educational experiences of the staff and students. This book shows how this commercial focus, while retaining the essential research ethos, is the underpinning feature of a leading University.

Professor dr. Ray Winger
Food science and technology
Institute of Food Nutrition and Human Health
Massey University
New Zealand

Preface to the first edition

This book originates from our experiences with the Product and Process Design course, which is one of the larger courses in the MSc Food Technology curriculum at Wageningen University. This course has been given now for almost 10 years and it has gradually evolved into its present form. In this course, students work on an assignment from a food company. The company supplies the students with marketing settings and the company's strategy and asks the students to design a product and process to make that product. Most of the work is desktop work, that is, students design the product and the process on paper and there is no time left to actually go to the manufacturing phase. Nevertheless, the industry setting makes it all very realistic and there is frequent communication with the company about the progress, which provides great feedback to the students. To help the students with this desktop work, some lectures are given to support them in the design phase. We couldn't find an existing scientific book to help them further at the MSc level, and therefore we decided to write our own.

Food product development consists of many stages. One of these is the design phase and it is this phase that the book tries to cover. Other equally important stages, such as finding out consumer wishes and needs, cost aspects, company and brand constraints, are not discussed in detail. More specifically, this book aims at supporting the food technologist in his/her design work. The basic message that we are trying to convey is that the technological phase of product and process design needs to integrate the relevant physical, chemical, biochemical and microbial aspects of food science and process engineering. It has, of course, a Wageningen flavour to it. Nevertheless, we hope that this book will also be useful for other universities and colleges teaching food product and process design.

We would like to thank all the teachers that have shaped the Product and Process Design course over the years into its present form; some of them are co-authors of this book. Furthermore, we would like to thank the publisher, Wageningen Academic Publishers, for their continuous support in making this happen. Finally, we gratefully acknowledge the contribution by Prof. Ray Winger from Massey University, New Zealand, who was so kind to review the draft of the book. His critical remarks helped us considerably in finalizing the book. Any remaining errors are of course our responsibility, and we would certainly welcome remarks and criticism from the reader.

The editors
Anita Linnemann and Tiny van Boekel
Wageningen, March 2007

Preface to the second, revised edition

After 4 years, the book was sold out. We decided to take the opportunity to update the book, and this has resulted in the second edition of Food product design: An integrated approach. The editors of the first edition are very happy that their colleague Karin Schroën was willing to join them in this endeavour. We have updated most of the chapters and included a new chapter on emulsions. The book continues to be used in the Product and Process Design course at Wageningen University and we hope that it will be useful for similar courses elsewhere in the world.

The editors
Anita Linnemann, Karin Schroën and Tiny van Boekel
Wageningen, March 2011

1. The need for food product design

Martinus A.J.S. van Boekel and Anita Linnemann
Product Design and Quality Management Group, Wageningen University

1.1. Introduction

The Odyssey, the second work of Western literature, tells us of the return of Odysseus from Troy to reclaim his threatened home on Ithaca. During his journey Odysseus lands on the island of the nymph Calypso. Calypso takes Odysseus to her cavern where Odysseus seats himself. Then, 'The nymph placed at his side the various kinds of food and drink that mortal men consume, and sat down facing the noble Odysseus. Her maids set ambrosia and nectar beside her, and the two helped themselves to the meal spread before them (Book V, 195-200)'. This extract from the Odyssey shows that gods consume just *one* type of food, namely ambrosia. This food is apparently all they need, and all they want. However, things are different for us mere mortals. We consume a variety of foods, and not a single one of them is sufficient to satisfy our demands in terms of taste or nutrition. This simple truth forms the basis of food product and process design.

The food industry has come a long way in food product development. The first large-scale activity was the preservation of food, so as to have food available outside the harvesting season. It involved activities such as pasteurisation and sterilization but also the scaling-up of age-old fermentation processes and drying techniques. Some decades ago, food production gradually became more uniform. This development took place when production processes changed from traditional methods to mechanized, large-scale industrial processing. Similarities in the needs of consumers formed the basis of the success of the products that the newly developed industries offered. In the course of time, industries started to diversify their supply of products. This diversification was stimulated in particular by the increasing prosperity after the Second World War in many western countries and the enormous effort put in increasing agricultural production. Advances in technology created the possibility to produce an ever-increasing number of diverse products, without violating the productivity of the industrial organisation. Along with increasing agricultural production, more attention was gradually paid to the nutritional quality of foods, and now, in countries with an abundant food supply, the food industry has changed from supply-based activities to large-scale, demand-based industries.

At present, the process of food product design is divided into several stages, namely (Earle and Earle, 1999):
stage 1: Product strategy development;
stage 2: Product design and process development;
stage 3: Product commercialisation;
stage 4: Product launch and evaluation.

This book is mainly concerned with the second stage of product design and process development, and in particular with those aspects that are the domain of the food technologist.

Figure 1.1 illustrates how the production of food is a combined effort of the producers of the raw materials, e.g. farmers and market gardeners, and the food industry. The way in which the transformation process from raw materials into food products is organized, depends largely on the information provided by the specialists in marketing and consumer behaviour. Together they outline the requirements of the present-day consumer and unravel the product features that would meet their expectations. This book does not deal extensively with that part of the new product design process, because this would require a book of its own, though some chapters touch upon this aspect as well. Useful references are Frewer and van Trijp (2006) and Moskowitz et al. (2009).

This introductory chapter, Chapter 1, describes why, in the context of developments in society and technology, food product design is necessary. Chapter 2 gives guidelines for structuring the innovation process by explaining how to effectively apply creativity in food product design. The following chapter, Chapter 3, presents a methodology to elaborate ideas for innovation into practical steps along the food chain, based on so-called Quality Function Deployment. Chapter 4 discusses the possibilities of modelling key reactions in the food product design process, Chapter 5 presents an overview on the way in which barrier technology can be used in the design of foods. Likewise, Chapter 6 discusses the possibilities of designing food emulsions. Chapter 7 deals with packaging design. Chapter 8 summarizes the most important aspects of hygienic design and Chapter 9 describes a method to assess the environmental impact of the production of a food, a topic in food product design that is considered more and more important by consumers as well as by governments.

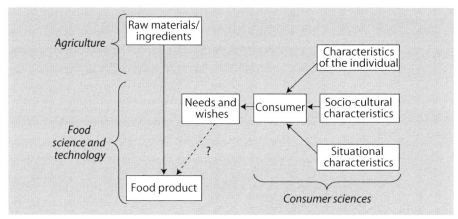

Figure 1.1. Relationships between agriculture, food science and technology, and consumer sciences in the process of consumer-orientated food product development.

Chapter 10 is about knowledge management and how this can be employed to support the product design process. The book ends with Chapter 11, which consists of an illustrative case interrelating many of the aspects from the previous chapters.

1.2. Societal setting

1.2.1. Chain reversal

Traditionally, the food market consisted of products that agricultural producers and food processors wanted to sell. However, in the past century, markets in developed countries have become saturated due to increased agricultural yields after the introduction of mechanisation and various agricultural inputs, such as fertilizers and all sorts of pesticides. This resulted in a change from supply-based activities to demand-based industries, which required product designers to reverse their mode of operation from supply-orientated into demand-orientated (Linnemann *et al.*, 1999). This process is known as 'chain reversal', and it implies that nowadays effective sales require a consumer-oriented approach. In other words, only a product that satisfies the demands of a consumer will be successful in the market. The implementation of such a consumer-oriented approach requires knowledge of the characteristics of the present societal setting as well as insight into the motives that drive consumers in their decision-making process while shopping for their food products.

1.2.2. Changing needs

Particular hurdles in satisfying consumers' demands and wishes are due to rapid demographical changes, which cause the fragmentation and diversification of households. Important demographical developments include an increased influence of ethnic groups in western societies and a shift in the age profile of consumers. Food products that fulfil the needs of the elderly have different characteristics than food products that are intended for young people, for instance, because of different sensory preferences and nutritional requirements. Another explanation for changing needs lies in the way in which lifestyles change over time. In the past, most people had jobs that required physical labour, whereas today many people work in offices behind a computer, implying that their required energy intake is much reduced. Moreover, many people prefer to spend their time on something else than the preparation of a meal, resulting in an ever increasing demand for food that can be prepared easily and fast. Finally, new scientific knowledge may lead to changing needs; relationships between food intake and certain health conditions become known, causing the need for foods that are composed in a new way. The ultimate implication of the apparent fragmentation was nicely expressed by a director of a Dutch supermarket chain when he said that at present there are 15, 16, or 17 million markets in the Netherlands, depending on the number of independent individuals in the country (Andreae, 1995).

1.2.3. Broadening of the quality concept

Consumers buy and consume products for a number of reasons. These reasons relate predominantly to the characteristics of the product, which should fulfil the consumers' needs as described in the previous section. However, the reasons to buy are also influenced by, for instance, the production methods that were used to obtain the raw materials. Therefore we divide quality attributes into so-called intrinsic quality determinants and extrinsic quality determinants (Van Boekel, 2008). The integration of the intrinsic and extrinsic factors finally determines the attractiveness of products to consumers.

The intrinsic factors refer to physical product characteristics such as the presence of taste and flavour substances, texture and shelf life. These intrinsic factors can be measured in an objective manner. Take, for example, the texture of a product. Texture can be defined in a physical-chemical way in terms of the composition of cell wall material and structure. Consumers describe these attributes with words such as crisp, mealy and tough. In addition, there are several other attributes like nutritional value, shelf life, freshness, safety and appearance. The combination of all these attributes together determines the intrinsic product quality.

Extrinsic factors relate to the way in which the food was produced, e.g. the use of pesticides, the absence of child labour, fair trade regulations, animal-friendliness, the type of packaging material, a specific processing technology or the use of genetically modified organisms during the production of ingredients. These extrinsic factors have no direct influence on the characteristics of the product, but they can be of overriding importance in the purchasing policy of consumers.

1.2.4. Globalisation

In recent decades, the world has witnessed an increasing integration of developing-country firms with geographically dispersed supply networks or commodity chains. These chains link together producers, traders and processors in developing countries with retailers and consumers in developed countries (Gereffi and Korzeniewicz, 1994). Liberalisation of global trade is increasingly accompanied by technical measures that impose quality standards regarding residues, additives and microbiological contamination. International sourcing of perishable products to secure year-around supply (under own label) is guaranteed through partnerships and long-term contracts. Retailers are devoting more and more shelf space to convenient high-quality fresh products (self-service) that are crucial for attracting and retaining wealthier customers (Marsden and Wrigley, 1996). Globalisation offers new opportunities and new challenges for product design and process development.

1.2.5. Chain approach

Finally, to efficiently and effectively incorporate consumers' wishes a chain approach proved necessary (Jongen *et al.*, 1999). In a food production chain, many actors may play a role: plant breeders, farmers, distributors, processors, marketers, retailers and consumers. In a consumer-orientated approach new product development will start with consumer and market research to identify the specific characteristics that a new food product has to have. The next step is to achieve co-operation and information exchange among all the actors in the production chain. This gives rise to several new issues, such as how the descriptive and qualitative terminology in which consumers express themselves can be translated into technological specifications. Followed by the question as to how the technological specifications can effectively be passed on to the relevant actors in the chain.

1.3. The role of science and technology

A great deal of knowledge about the properties of foods has accumulated. In that sense, food product development has changed from an empirical approach ('changing recipes') to a systematic science-based activity. In other words, scientific knowledge is of great help in successful food product development. It combines knowledge of raw materials, food products, process engineering, marketing and consumer research. Food product design can in this respect be seen as a technological activity based on science, engineering and social sciences. It is characterised by a high level of integration of these scientific activities. It is important though to realize that consumers are not aware of this, and may not even want to know about all this. Consumers only see the results of these activities, namely the food products that are offered to them. Figure 1.2 illustrates this. At the core of food product design are the scientific disciplines of food chemistry, food physics, food microbiology, nutrition (nutrigenomics), and toxicology, which form the knowledge base for food product design, together with process engineering. Product and process design are closely linked.

In order to understand foods, simplified model systems are often used that mimic real foods to some extent. In this way, matrix effects and interactions between components are taken into account. A nice example is the Maillard reaction that is so incredibly complex in real foods that it is virtually impossible to study. Instead, aqueous solutions of sugars and amino acids are studied, from which basic rules can be derived that can be applied to foods. This then already requires some integration of the basic disciplines, and processing also comes into play. One step further and we are in the domain of food product and process design. It is essential to realise that product and process design go together: they are two sides of the same coin. Innovation in food production can be brought about by changing food composition and structure but also by novel processing techniques. Furthermore, the choice for a certain product composition has an effect on the process that is needed to

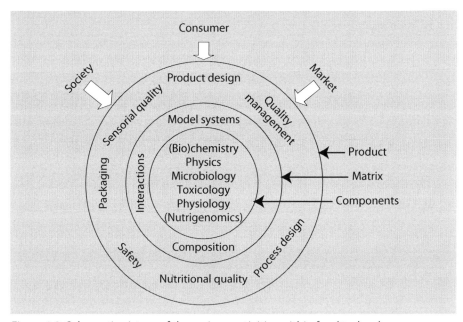

Figure 1.2. Schematic picture of the various activities within food technology.

produce the food. For instance, if the product is solid, use of a continuous flow heat exchanger is going to be very difficult. On the other hand, the applied process can have a profound effect on the composition of the resulting product. Furthermore, as indicated in Figure 1.2, the third layer of the model also indicates safety, nutritional and sensorial aspects, indicating that these are properties that are essential to consumers. Quality needs to be designed into the product and in order to achieve the required level, quality management is of the utmost importance (this can be achieved with management systems such as HACCP). A useful reference in this respect is Luning and Marcelis (2009). Then, of course, the surroundings are important, trends in society have an important influence on food choice and the way consumers look at food production.

1.4. Developments in food processing

As stated earlier, product design and process design go hand in hand. There have been changes too in the field of processing. New preservation technologies have become available and this has opened up new possibilities for new concepts in food products. One of the most striking of consumer wishes is the desire for fresh-like foods that also have a long shelf life, for instance in the form of ready-to-eat meals. With new technologies such wishes may be within reach. The phrase 'new technologies' implies a comparison with 'old technologies'. The unit operation of comparison is heating, which is traditionally the oldest preservation technique and is still widely used. Recently, there has been more interest in non-thermal preservation techniques because the unavoidable damage caused by heating conflicts with the consumer's desire for fresh-like foods. Table 1.1 gives an overview of possible new techniques. There is currently a lot of research into high-pressure techniques, pulsed electric fields, and ohmic heating. It is fair to state though that these techniques are not yet being widely applied. All the same, developments in technology also must be monitored closely by food product developers to assure that they always make optimal use of the possibilities when designing new food products.

1.5. Tackling consumer preferences

The most important consequence of the above described 'chain reversal paradigm' is that consumer wishes are the starting point for product and process design, However, consumer behaviour has become increasingly unpredictable because of the many factors involved, as described above. Marketing specialists describe the present consumer as 'capricious, impulsive, spontaneous and irrational'. This description reflects the frustration about the inadequacy of many laboriously elaborated marketing strategies. Meulenberg (1996) analysed the Western consumer and his buying behaviour in relation to social, demographic and economic circumstances. The choices made by present and future consumers will primarily be directed by societal developments, assuming that there is still an abundance of food products and that the supply of basic needs is easily provided. Meulenberg (1996) described so-called pure lines of present and future consumer types. The advantage of this approach is that such generalized consumer types can be linked to specific product characteristics (Table 1.2). Naturally, pure consumer lines are an abstraction of reality and do not exist as such, since individuals express a combination of attitudes. Such attitudes are influenced by many factors, including consumption situation and consumption time (Sijtsema, 2003). In general, certain consumer lines will become more important and others will decline in time, in response to societal and/or economic developments.

1.6. Concluding remark

This book attempts to present the state of the art concerning the important technological matters that come into play for appropriate food product and process design. The subtitle "an integrated approach" reflects an important philosophy: it is the editors' conviction that successful product and process design truly requires integration of several scientific disciplines. This is an easy statement that is, however, not easily brought into practice. It is hoped that this book sets the scene for such an approach.

Table 1.1. Overview of new developments in food technology (adapted from Van Boekel, 1998).

Type of technology	Developments	Characteristics	Suitable for	Drawbacks
Preservation by thermal treatment	Htst, uht	Less heat damage, heat regeneration	Fluids, fluids containing particulates	Not suitable for 'solid' foods
	Ohmic heating	Rapid and even heating	Fluids and fluids containing particulates. Low-acid foods	No heat regeneration
	Infrared heating	Efficient and rapid heat transfer	Baking, drying, grilling, thawing	No heat regeneration, low penetration depth
	Microwave	Rapid heating	All kinds of foods	Uneven heating, no heat regeneration
	Sous-vide	Mild heating in vacuumed pouches, sensoric quality improved	All kinds of foods, catering	Shelf life limited
Preservation by non-thermal treatment	High-pressure	No thermal damage, fresh-like quality	Fruits, vegetables	Not continuous, expensive
	High-electric field pulses	Little thermal damage, high sensoric quality	Conductive foods	No inactivation of spores, high cost
	Pulsed light	No thermal damage	Packages	Only active at surfaces, or in transparent liquids
	γ-irradiation	No thermal damage	Fruits, vegetables, spices, condiments, meat	Not accepted by consumer
	Ultrasound	Facilitates heat and mass transfer, physical disruption	To be used in combination with other techniques	High cost, little microbial stabilization
	Drying, concentration	Removal of water, glassy foods	All kinds of foods	
	Glassy foods	Solid amorphous materials, very stable below glass transition temperature	Low moisture foods (confectionery, cereals, powders, frozen products)	Stability critically dependent on glass transition temperature
Preservation in the cold	Rapid freezing	Formation of glass	All kinds of foods	
Hurdle technology and minimal processing	Combination of preservation technologies	Less quality loss	Dry meat products, fruits, vegetables	Microbial safety is critical

Type of technology	Developments	Characteristics	Suitable for	Drawbacks
Packaging	Aseptic packaging	Combination with continuous processes	Fluids, fluids containing particulates	Special hygienic design of equipment
	Modified atmosphere (ma) and controlled atmosphere packaging (ca)	Interference with metabolic processes of the food	Fruits, vegetables, meat, seafood	
Fermentation	Continuous processes	Higher production rates, standardized quality	Milk products, alcoholic beverages	
	Solid-substrate	Better control of fermentation	Solids, such as soy beans	
Separation	Membranes	Separation of ingredients on large scale	Liquid foods	Fouling of membranes
	Chromatography	Separation of ingredients in high purity	Liquids	Expensive
	Enzymatic synthesis	High purity	Peptides, colorants, emulsifiers, flavours	Expensive
	Supercritical fluid extraction	Efficient extraction of ingredients	Coffee beans, hops, oils and fats	High cost
	Reversed micelles	Protein extraction	Novel protein foods	
Fabricated foods	Extrusion	Shaping and or cooking	Snacks, pasta, pastry, novel protein foods	
	Emulsification	Emulsion droplets contribute to structure, flavour, nutritional value	Fat-containing products	
	Phase separation	Structure formation by food hydrocolloids	Gel-like foods, novel protein foods, meat and fish-like products	

Table 1.2. Future consumer prototypes, some key characteristics with respect to food preferences in general and, as an illustration, to rice products in particular (adapted from Linnemann et al., 1999).

Consumer prototype	Characteristic food preferences	Characteristic preferences with respect to rice products
The environment-conscious consumer	Organic foods Unprocessed (fresh) foods Foods from short production chains	Organic rice
The nature and animal-loving consumer	Foods without genetic modification Animal-friendly produced meat	Rice without genetic modification
The socially-conscious consumer	Foods from fair trade	Fair-trade rice
The health-conscious consumer	Foods with health-protecting and health-promoting properties (e.g. low calorie, with extra vitamins, minerals, and non-nutrients)	Fortified rice, brown rice
The convenience consumer	Ready-to-eat meals Fast food Take-out meals Restaurant food	Ready-to-eat rice (e.g. in pouches for the microwave, in cans)
The hedonic consumer	(Exotic) specialities, delicacies Foods of high sensory quality	Exclusive rice varieties
The price-conscious consumer	Homemade meals, with ingredients of a favourable price/quality ratio (e.g. products from large-scale production)	Low-priced rice
The variety-seeking consumer	Seeks diversity in ingredients for meals New product introductions	New rice products and all sorts of combinations with rice, e.g. differently seasoned rice

References

Andreae, J.G. (1995). 15 miljoen markten [15 million markets]. Conferentie Massa-Individualisering. WTC Rotterdam. 24 oktober 1995.

Earle, M. and Earle, R. (1999) Creating New Foods. The Product Developer's Guide. Oxford: Chandos Publishing Ltd.

Frewer, L. and H. van Trijp, eds. (2006). Understanding consumers of food products. Woodhead Publishing, Cambridge.

Gereffi, G. and M. Korzeniewitz, eds. (1994) Commodity Chains and Global Capitalism. Westport, Conn, Praeger.

Jongen, W.M.F., A.R. Linnemann en M. Dekker (1999). Product- en procestechnologie [Product and proces technology]. In: Werkende ketens. Keesing Noordervliet, Houten. pp. 28-34.

Linnemann, A.R., G. Meerdink, M.T.G. Meulenberg and W.M.F. Jongen (1999). Consumer-oriented technology development. Trends in Food Science and Technology 9: 409-414.

Luning, P.A. and Marcelis, W.J. (2009). Food quality management : technological and managerial principles and practices. Wageningen : Wageningen Academic Publishers

Marsden, T. and N. Wrigley (1996). Retailing, the food system and the regulatory state. In: N. Wrigley & N. Lowe (eds) Retailing, Consumption and Capital: towards the new retail; geography. Harlow, Longman. pp. 33-47.

Meulenberg, M.T.G. (1996). De levensmiddelenconsument van de toekomst [The food consumer in the future]. In: Markt en consument (NRLO-Rapport 96/4). Part 1. NRLO, The Hague.

Moskowitz, H.R., Saguy, I.S.. and Straus, T. (2009) (eds.). An integrated approach to new food product development. CRC/Taylor & Francis, Boca Raton, FL.

Sijtsema, S. (2003). Your health!? Transforming health perception into food product characteristics in consumer-oriented product design. PhD Thesis. Product Design and Quality Management Group, Wageningen University.

Van Boekel, M.A.J.S. (2008). Kinetic modeling of reactions in foods, CRC/Taylor & Francis, Boca Raton, FL.

2. Creativity and innovation

Corrinne J. Goenee
White Tree B.V.

2.1. Introduction

Innovation and business development lie at the very heart of success. The words signify the discovery of new opportunities and solutions based on in-house competencies, content, know-how and skills. Pro-active, constantly looking for new ideas, with the viewfinder fixed on the future. Innovation is a keyword in contemporary society, as is creativity. But what do we really mean when using these words, and how are they related to each other?

For a long time, the focus within organisations and education programmes has been on selection procedures and routines, how to get things done and implemented, whereas today's focus is shifting towards the beginning of the innovation process: how to find that brilliant idea we are all looking for. We have somehow lost the hang of this, because our minds are used to focusing on increasing productivity and/or reducing cost. How then do we reinvigorate this innovatory spirit? Do we merely need to establish an infrastructure to pick up ideas from an ongoing train of thoughts, or should we try high-pressure brainstorming? Or should we instead remodel or refuel the existing company's culture and values in order to give ideas a chance to grow before they are lost in the chaos of everyday activities and day-to-day troubleshooting? Both elements, procedures and culture, are crucial to successful innovation, but before you even get started on the hunt for ideas it is worth establishing your guidelines. What do you have already and what do you want to achieve?

Company, brand and product positioning is a key success factor in this context. What do you represent in the midst of your competitors, and how does your product stand out from other products? What do consumers get to see in the end? Innovation is not the easiest or most standard path for a company to follow. There are many potential roadblocks, which we will highlight. After that we will look at ways to increase your own and your company's 'innovativity' and that of the company you are working for.

2.1.1. Context (roadmap to innovation)

We believe innovation needs to move beyond the procedural, rational framework that guides the process. Of course this is a necessary element for a job well done, and an important stepping-stone for providing the focus needed to deliver top results. But the real difference is made when people and teams are able to broaden their perspectives, see new opportunities and learn from each other and the world around them. This chapter is about the magic of our minds. About how intuition and creativity influence innovation processes. In order to understand the role of creativity in the innovation process, we will start by highlighting the major steps in the innovation/NPD process.

Market knowledge

First, it is crucial to know the market one is in. Usually, research is performed or figures are bought, in order to understand how the company (its brands and products) performs, and how consumers perceive the brand/products. Creative research leads to surprising views on perceived 'borders and gaps' in the market place.

Company knowledge

Second, one needs to know or establish what one can and/or wishes to offer and how much one is willing to invest in order to achieve a (growth) goal. A creative perspective brings a broader perspective to your possibilities.

Strategy

Third, one needs to choose direction. Usually, this is done based on figures (which segments are growing) and trends (consumer needs). However, intuition and creativity (for us, intuition is a body of knowledge you are unaware of ever acquiring and which cannot easily be expressed through language) helps people to 'see' what no one else has seen, to believe in an opportunity when others don't (yet), and act on it.

- The Health Trend was hard to miss, but within the health focus, the persistent roll-out of probiotics, first by Yakult, then by other dairy companies, such as Danone, was based on a strong vision and determination.

Tactics

Fourth, establishing a strategy and opportunities is one thing, responding to them is another. It is not at all easy to bridge the gap between a trend and a product, especially since one needs to adhere to existing (brand) strategies and fit into existing markets, yet differentiate oneself from existing brands/products. Creativity comes into play at this stage for discovering ways to present new and surprising responses to given opportunities, which are still recognised and valued by consumers. It is not enough for the idea (future product) to be 'nice', because consumers will also have to be willing to pay a certain price for it. The well-known term 'brainstorming' usually refers to this step in the process.

- Danone made some choices when working on probiotics. These choices were made before a final product and packaging design were established. Instead of focusing on the ingredient (bacteria), focus is on the action/benefit of the product. Actimel aims at supporting natural defences (resistance), Activia claims to relieve digestive discomfort.

- Diet colas attracted and were initially targeted at young women. Then Pepsi decided to introduce a low cal version of its cola (= health sub trend), especially for men. This product was introduced by the name 'Pepsi Max' (Maximum Taste, No Sugar).

Execution

Once an appropriate goal and 'message' have been formulated, work must start on the actual product, packaging, design, advertising plans, which reflect the intentions formulated in the briefings. This is the kind of creativity which is well known (spontaneously associated with advertising agencies). Creativity is the ability to provide answers, in tangible products, to the needs in consumers' minds. However, at least as much creativity is needed for solving technological and/or production problems. This kind of 'creativity' is usually overlooked.

- Hero 'Fruit2Day'. The bottle reflects 2 pieces of fruits, thus reinforcing the message of the concept: this drink delivers 2 pieces of fruit (daily recommendation) in one bottle, the easy way to take care of your health.
- Knorr Vie is doing the same thing, but in this case the bottle is really small, thus enhancing the 'power' of the product.
- Vla-Flip is an introduction of the dairy company Mona. The product is similar to a well-known traditional Dutch dessert in the sixties, and consists of (sour) yoghurt, (sweet) vanilla custard, and some fruit jam. There was a problem with industrial production though. A mix of Yoghurt and Custard is not stable. The solution product developers came up with consists of using only one mix for both components, with a colour and more sugar added to one part of it (the custard).

Now that we have established the context, let's go on to talk about what creativity is and how it can be enhanced, in companies and on an individual basis.

Innovation is a must, but it doesn't just 'happen'. Due to developments in our society, changing values, priorities in business (cult of efficiency) and the current rate of innovation, it is difficult to live up to the expectations. Inherently, innovation is a tough job for any company.

New Product Development (NPD) involves going against the flow. The company is meant to produce within economic guidelines. This means that every organisation is working on the rationalisation and optimisation of existing processes (more of the same). Innovation requires a mental approach that is the exact opposite.

Innovation relies on push marketing. Consumers react and validate, but they are not the ones who bring you fine-tuned concepts. Market research may present you with areas of opportunity and restrictive conditions, but it does not offer ready-made solutions or concepts! It takes time and effort to discover the consumer's preferences and 'hot-keys', and to figure out an appropriate way of responding. The ultimate challenge for you is to find out what consumers want before they know it themselves.

NPD is not an exact science like mathematics or physics. Innovation, unlike audits or re-engineering, is not given to formulas. While it can be supported by systems, it can never be reduced to systems. Experience is a valuable help.

There is no guarantee of success. NPD involves risky decisions. The outcome is uncertain. If your job is on the line, you do think twice! But there is no guarantee of success. Market research helps, of course, but we also know that consumers don't always do what they say they will do. And you cannot wait too long, otherwise your competitor might take your market share. As a Formula 1 driver once said: 'If you have everything under control, you are simply not moving fast enough!' Often, after a successful introduction, we hear marketers and developers claim that they had a similar product on their shelves, but never introduced it. So it takes a convinced mind to make decisions. It takes a lot of courage and persistence to persuade management to invest time and money in any project. Many projects have been shelved because there was nobody championing them, and nobody convinced enough to take the project through to a successful conclusion.

2.2. Key success factors for innovation

There is a big difference between 'creation' and 'innovation'. Creation means thinking new things. The thought, the idea, is what counts. The purpose of creation is a means in itself. Innovation means doing new things with a commercial purpose. It means getting things done (for which you need perseverance skills). A new product or a new process (for better products) has to be initialised, developed and put into place. Nowadays companies find themselves on different rungs of the 'innovation ladder'. The key is that there is an awareness of the necessity for innovation, and there are several ways to stimulate people to spend time and effort on the process. Ideally there is a continuous flow of ideas and people take independent action to get things done. But this is easier said than done, especially for more ambitious ideas. Nonetheless, many companies are actively concerned with innovation and are improving their innovative strengths, more or less in line with the following stages of development:
1. 'We need to innovate'. Management has decided that it is important for us to do so.
2. Create conducive conditions (= organisation). Management communicates that creative and innovative thinking is to be encouraged. Sometimes they offer rewards for innovative ideas, or they find other ways to stimulate employees to think creatively and come up with viable ideas for product development or (more) efficient or cost-cutting production techniques.
3. Innovation teams (= organisation). Management has given a specific assignment to a selected team of individuals to come up with new projects (usually when the first two stages have not proven successful on their own). Usually management has also allocated resources (time and money).
4. Processes: idea-concept-project. The team, or department, is trained in how to handle new projects. Usually, new projects are different, somewhat foggy, and

the developers cannot build on previous experience. The projects do not fit into everyday routines, and therefore demand special attention. Once the new process is in place, time can be won in bringing the idea to the market. But while it helps to have an infrastructure, without further efforts (no. 5), nothing will happen. The tracks are in place, but do not think trains will start running just like that.

5. Tools (= activation). For a train to start running, someone has to take action. There are several things that need to be done. Strategic marketing research is performed, to determine 'gaps', consumer research to search for opportunities and guided brainstorms to fill the 'basket' with ideas. Concepts are being developed to show strengths and make them appealing both to internal decision-making managers and consumers.

6. Continuous 'process' (= organisation and activation and flow). Ideally, there is a continuous flow of ideas and people take independent action to get things done. This is a much envied ideal, which occurs in some companies, sometimes. This situation seems easy and logical, but in reality, is hard to obtain.

It is not sufficient to have processes in place and information available. It's the people who will have to make 'it' happen. When it comes to innovation, sharing, learning and cooperating are the main ingredients for success. This can only be achieved when there is a high level of trust and confidence within the company.

Innovation occurs in the boundaries between mindsets, not within provincial territory of one knowledge and skill base (Leonard-Barton, 2005).

'The process of encouraging innovators is never-ending... and your actions must reflect your words. There's an American slang term that describes this perfectly: You can't just talk the talk... you've got to walk the talk. Listening is a big part of the job of encouraging innovation. Pay attention to every idea... no matter how unlikely. Today's loser might become tomorrow's winner. For example... some of our people invented a material that absorbed oily liquids. We sold it as a cooking aid. Drop a square of this material into soup... and pick up most of the unwanted fat. For some reason... this product just couldn't get going. 'However... today... we manufacture a product with similar characteristics. This time around ... people use this material to absorb oil spills at sea and chemical spills in factories.' (From a speech given by Dr. M. George Allen, 3M)

2.3. Organisational requirements for successful innovation projects

Analysis conducted by a major chemical company into the origin of its successful products revealed that the majority could be traced back to three or four people. These people were researchers, with a relatively low level of education and a low rank within the organisation. Their skill was being able to play with chemicals and develop ideas in an environment where their ideas were taken seriously (Henley Centre).

Rosabeth Moss Kanter studied a couple of highly innovative organisations and came to the conclusion that what innovative companies have in common are: Structure and Culture, in other words routing (knowing where to go) and motivation (readiness to readjust). As regards the structure, it helps to have systems and procedures in place. It helps to know where to go with a good idea, it helps to install a committee of some kind which devotes time to the stimulation of innovative thinking, and it helps to get people together regularly for thoughtful discussion and idea generation. However, it requires a 'receptive' culture to actually keep the train (of thoughts and actions) running on the tracks, and to make projects out of concepts. We know that people need a safe place to experiment, they need support and rewards (not necessarily financial rewards – appreciation and/or status are often more important), and they need short communication lines and ideally some empowerment to take decisions on their own. If all that is secured then there is a good innovation climate, and people will be encouraged to think new thoughts.

Some people have a natural tendency to challenge current situations and improve on them. These people are known as 'natural innovators'. For them, merely improving the 'climate' may be enough, but they are few in number and they do not necessarily have the knowledge or communication skills needed to translate ideas into action. That is why more effort is now being put into communication between departments, into team building, and into developing innovative skills in more employees. Somehow it seems strange that in today's cultural and educational systems so much attention is devoted to analysis, dissection, deduction, conclusion and so on, while so little attention is paid to 'the other side of the brain'. It seems that creativity is regarded as a talent, like being able to draw. It is odd that if you enter an art classroom and draw something then you are immediately told that 'you have a talent' or that 'you will never be any good at this'. In a maths class, on the other hand, you are taught the principles regardless of your talent. 'Talent' can grow; it can be developed. If only the truly gifted were to receive maths education, then those classrooms would be fairly empty.

Creativity is as essential a skill as is the power of analysis. Consider the next section on creative thinking techniques as a guideline to how your mind really works. It will give you a start in discovering your genius. You don't have to become the next Da Vinci or Einstein. Just try to 'optimize the magic of your mind' (Parnes, 1997), and direct it at your assignments and jobs.

> *Creativity is the connecting and rearranging of knowledge –in the minds of people who will allow themselves to think flexibly- to generate new, often surprising ideas that others judge to be useful (Plsek, 1997).*

N.B. Again, for business innovation, creativity alone is not enough. Sometimes, as in the case of food innovation, it is best to have a multidisciplinary team working on innovation after ideas have been generated and in the concept development phase, so that specific knowledge surfaces at the right time and in the right place. Effective communication on abstract or intangible matters or on 'not yet existing' products at the start of the innovation process shortens briefing time and project efficiency considerably. A side effect of communicating or brainstorming with specialists from several departments (marketing, R&D, production and packaging) is that everybody is forced to make their ideas or knowledge 'simple' in order to make non-specialists understand. As Einstein said: 'If you can't explain it to your grandmother, then you don't really understand it (yourself)'. This also prevents the use of jargon, so that specialists cannot hide behind words and accepted 'truths' but are forced out of their habitual routines and assumptions. All of this contributes to better communication and liberates free 'creative' thinking. This is why we sometimes call creativity 'shared imagination'.

2.4. Creative thinking techniques

Much NPD work follows a well-worn path of market research, segmentation, competitive analysis and forecasting before fixing on the consumer opportunity and passing the brief to R&D. I would argue that, for two reasons, these techniques should only be used to illuminate rather than determine the creative process. First, these techniques are used by everyone - a commodity available to anyone at a price, unlikely to generate long-term competitive advantage. Second, they cannot in themselves generate new ideas. The tight brief they create can constrain instead of aid the creative process. The resultant products often provide incremental advantages over previous products, but with a short-lived competitive advantage. A strategy of letting idea-generators have their head increases the chance of failure, but also increases the likelihood of a major success with a sustainable competitive advantage (Jon Francis).

Everyone is gradually starting to grasp the value of creative thinking. Especially in today's society it is evident that analytical thinking is not enough to retain one's competitive advantage or to offer customers a unique product or service. Innovators are considered to 'be' creative. We believe that 'creative thinking' is also a mindset, more than just a mere fluke. Creative thinking is a question of seeing things in a broad perspective, of thinking freely. There are tools and processes that stimulate thinking without limitations, and at least offer a means of more easily steering the thinking process, so that a 'flash' of inspiration can also be prompted when it is most useful to you.

The method is ideal for companies that don't want to leave anything to chance, but prefer to find products that answer existing and future customer demand in a structured way. Make use of what you already have. Capitalise on 'the image in the consumer's head'. Make product launches logical and credible, make products more appealing and briefings more inspiring, whether the emphasis is on deciding a company direction and strategy, on communications or on brand, product and/or packaging development.

2.5. Key success factors for business creativity

What we need is:
• distance;
• direction;
• focus;
• inspiration;
• communication.

2.5.1. Distance

We focus on what we know and recognise what is familiar. We match that with our memories and assume the rest. Somehow we tend to automate our thinking. This is only natural, but when it comes to creative thinking this works against us. Most creative people have one thing in common: they 'wonder'. They don't take anything for granted, but keep asking awkward questions.

In companies, it helps tremendously to take a step back and take a look from a distance, as if observing what is happening from a balcony. There are several techniques to help you establish this distance. There are also a multitude of business games that can help us to establish that distance and can be used in combination with thinking techniques. And of course, establishing a literal distance, gathering at a location that is not your usual business environment, works wonders as well.

The fact is that distance is crucial in order to give your mind some space for 'freestyle thinking'. For 'mental play' it is essential in order to overcome your natural resistance to new projects, which never seem to fit in with existing routines, demand too much time, and never seem to align with your existing policies and strategies. It just doesn't come easy.

2.5.2. Direction

A long time ago we ascertained that mere 'brainstorming' (sitting down together and shouting out ideas) did not work as well as we would like. It only works when the group, the individual brainstormers, are highly creative, and/or when the purpose of the gathering is well defined (not 'let's think of new products', but 'let's solve this particular problem', or 'we want to be the first to launch this kind of product for that particular demand or target group'). The point is that the result of an 'unguided' brainstorm is highly splintered, and even if some good ideas came up then there is a danger that they could be buried under the load of information and ideas that were raised.

If you are aware of these conditions then you might use the initial flow of ideas in order to define your direction: 'These ideas are good, others are not so good, because....' 'This means, that we will have to dig deeper into this particular area of interest.' Also, it is human nature to drift away on opinions, soon losing sight of the original objective (finding new product ideas), instead of defending the pros and cons of ideas that came up earlier. It is therefore useful to define your goals and objectives before getting started, thus concentrating your efforts on effective communication.

It may be helpful to use some guidelines, such as a SWOT (Strengths and Weaknesses, Opportunities and Threats) analysis. After all, if you are aware of your strengths and weaknesses and if you are up-to-date with consumer research, current trends and future market developments, then you can analyse your situation and select the direction that most needs your attention. But always bear in mind that analysis alone will not lead to your desired concepts or concrete projects. It gives direction, but no straight answers.

In the end, it's all about responding to opportunities and threats, but it takes time to do it right. A lot of homework is involved, and a fresh approach is needed if you are to step away from existing mindsets in the company. 'The fish are the last to discover the water'. If you are right in the middle of a battle field, it is hard to see clearly.

2.5.3. Focus

Our experience is that the more focused the brainstorm, the better the results. This follows on directly from the previous comment. First you need direction and then you need focus. Try to avoid solving 101 problems in one go, because your brain simply cannot handle it. If we focus on a particular objective then we are bound to find the answer. By the way, when focusing your mental activity in this way you will use all the information that is foremost in your mind (and to some extent what is at the back of your mind, too) and seek out possible connections to that objective. We find that the most 'promising' ideas have one thing in common: they offer more than one benefit to both consumers and the company.

2.5.4. Inspiration

> Keep on the lookout for novel and interesting ideas that others have used successfully. Your idea has to be original only in its adaptation to the problem you are currently working on (Edison).

Inspiration is everywhere. You just need to open your eyes and actively look for it. If you deliberately try to mentally connect everything you see to the area of interest you are studying then you will find new ideas, new products and new applications.

In essence this was Da Vinci's trademark. He apparently took notes on everything, so we find a sketch of a wonderful sunset alongside the anatomy of a cat and a shopping list on one single page. It was his way of stimulating his mind, his own thinking.

Many creative thinking techniques are based on association (deliberately connecting things you see with your objective) and analogy (finding answers in other contexts than your own). You may want to walk into stores with specific assignments for yourself (look for really good or really poor examples of what you want to achieve), you may want to compare your situation with similar situations in other contexts (analogy), you may try to apply really extreme situations to your own sector, you may want to do virtual store checks (new product databases), or simply take a walk on a beach or in the woods. Talk to outsiders. Listen to what they are trying to say. So much is possible. Just make sure that your objective is clear, that your mind is pondering a particular issue, and then purposefully and actively translate everything you see, hear and feel back to your issue. You will discover new perspectives, new insights and new solutions, no doubt.

2.5.5. Communication

Human beings have a tendency to believe that 'what we see is the world as it is, and not the world as it is seen by us'. Communicating with others helps us to gain new insights, new flashes of inspiration. Ultimately that is how we learn. In addition to this communication (which does not have to be spoken, as books and internet do the job as well), we all know that discussing issues with others and explaining what we mean helps us to order our thoughts and clarify them. This principle is used for group brainstorming. Getting together in order to discuss a particular issue and find new perspectives, new insights and ideas for a problem or opportunity is useful in some of the stages of 'idea development'.

2.5.6. Brainstorming

The term 'brainstorming' has become a commonly used word in the English language as a generic term for creative thinking. The idea behind brainstorming is the generation of ideas in a group setting based on the principle of suspending judgment - a principle that scientific research has proven to be highly productive as an individual effort and as a group effort. The idea generation phase is separate from the judgemental phase of thinking.

In 'Serious Creativity', Edward de Bono describes brainstorming as a traditional approach to deliberate creative thinking, and people therefore think creative thinking can only be done in groups. The whole idea of brainstorming is that other people's remarks will serve to stimulate your own ideas in a kind of chain reaction of ideas. Another advantage is that if there is a good programme and discussion leader then effective sharing of knowledge takes place. Our conclusion is that groups can be helpful, but they are not essential for deliberate creative thinking and there are many techniques that individuals can employ to generate ideas.

If you decide to do a brainstorm, make sure you have:
- a clear objective;
- a solid basis (information) and focus: a point of departure and of arrival;
- an understanding of influencing factors, external (customers) and internal (company);
- a motivating source of inspiration;
- accessible and perspective-broadening creative thinking techniques;
- a budget and management support, to ensure that good ideas have a chance to become projects (when this fails then the motivation to contribute goes downhill fast).

2.6. Required mindset

Unfortunately, innovation, unlike audits or reengineering, is not given to formulas. It is given to people - restless, inspired, fascinated individuals with an almost cellular need for change. And while it can be supported by systems, it can never be reduced to systems. 'Innovation,' as Tom Peters so aptly put it, 'is a messy business.' If you want to spark innovation, forget about slick formulas for a minute and pay attention to what's happening on the inside. Because that's where it starts. With the innovator – the inspired individual who sees a better way and goes for it. And the key to the innovator? The special blend of inner qualities that allows him or her to succeed when others have long since gone home. Tools? Techniques? Five-step models? Sure, they're useful. But, without the user of them having the right stuff, they're merely decoration – not unlike having a new set of jumper cables, but no car (From the 'Free the Genie' series by Mitchell Ditkoff).

For individuals as well as groups there is one major condition for hitting the mark when it comes to thinking creatively: you need to get your mindset right. You need to be willing to actively and deliberately direct your thoughts towards your objective, and you need to allow yourself some space to think.

Don't rush to conclusions before you even get started. Take a step back and think. This, again, may sound easy, but we have a 'natural' (or acquired) tendency to jump into action and to judge both our own and other people's thoughts and ideas. To criticise them. This seems to be related to Platonic and Socratic influences: if we find all flaws, we will end up finding the truth. You need to bring that initial reaction to a halt for a moment. Allow yourself the freedom to conceptualise without (immediately) judging your ideas in terms of the real world. In his meetings, Osborn (third edition 1993, first edition 1953) established the rule that everybody would simply table their ideas before the coffee break. After the break they would start evaluating them. This turned out to be very effective, as it stops people following their natural tendencies and encourages them to discuss things constructively, building on each other's ideas, rather than defending their own. But the same rule also applies to you. Maybe some ideas that you would normally discard immediately have some spark, some interesting element that leads to better ideas. Some ideas may turn out to be the stepping-stone to better ideas, following the same principle. Stop looking for the golden egg, and try building a house with the bricks you find on the way to your objective. There is a metaphor we use for this phenomenon. If you don't let the bees out of their carefully constructed 'cocoons' then there is no way that they can fertilise each other. Think while relaxed, let your thoughts come and go. This is mental play! Anybody can do it. You can do it too, because creative thinking is based upon thinking capabilities everybody possesses! To quote Plsek (1997), 'If you can think, you can think creatively.'

2.7. Required thinking capabilities

The basic abilities that you need are observation, visualisation, imagination and association.

Observation is crucial. If you are aware of your own observations, if you know how to look at things, then you are well on the way to making discoveries. To give a visual analogy, there are many books on photography and design (see references). These deal with visual perceptions, based on principles and techniques that can easily be applied to your 'thinking' as well. The basic points that we can learn are to regard the situation as a whole (helicopter view), change your 'point of view' (literally), and zoom in on elements you find attractive (or problematic).

Visualisation is about being able to visualise situations and problems. We know it is a great help in solving problems. Anybody can visualise. Think of your Prime Minister. You probably have his or her face etched in your memory. It is easy, as you must have seen him plenty of times on TV or in newspapers.

Imagination is a true gift of nature, which lets us imagine things that either do not exist or are not actually present, or things that we have never seen before. It is essential for any kind of development, if you want to do more than simply copy your competitors. You need some fantasy, but that is also something we all have. As an exercise, think of your Prime Minister in swimming trunks. I bet you haven't seen that in a photo or video!

Association is pivotal as well, and it is something that everybody does all of the time. This is what your brains are good at. It is how they function and how they learn. There are many techniques that reinforce and develop this trait. The most illustrative are mind mapping techniques as described by Wycoff (1991).

Given the fact that everybody possesses these abilities, let us now focus on creative thinking techniques. Again, 'creative thinking' is a mindset. You can also make a deliberate effort to think creatively in business situations (somehow creativity doesn't seem too much of a problem in private situations; it is when we need to be 'professional' that we discard creative thinking as something not serious enough for business problems).

We can define a couple of steps in the process. Beginning creative thinkers (or 'swan thinkers', those who naturally find themselves sticking closely to the subject and goal) need to acquaint themselves with each individual step, while advanced creative thinkers (or 'butterfly thinkers', those who naturally find themselves mentally flapping around an issue) may find themselves skipping some of them. There is no right or wrong here. Just try to be consciously aware of what is happening in your mind and in the mind of others. Remember that creativity techniques can be used as a group activity or by individuals. First of all, try to understand what is happening, then try to apply them to a 'problem' you are facing.

2.8. The brainstorming process

1. Create time and space for problem solving.
2. State your problem, and then challenge your statement.
3. Find spontaneous solutions.
4. Then make use of 'creative thinking techniques'.
5. Always evaluate.
6. And make sure you pick up the pearls and build on them (concept development).

2.8.1. Time and space

Time and space is essential to creative thinking. For group work, you need to organize a time and place to get together. For individuals, you need to find out which place suits you best. Ideas usually come to people when they are doing something boring for their minds (e.g. ironing, mowing the lawn or marking hundreds of little stripes on a piece of paper), or something that does not require their full concentration (e.g. taking a shower or, dare I say, when driving a car on the highway). For you it might work to try the techniques when taking a walk, when sitting in a bathtub or on a summer terrace. However, it is crucial to remember that you are working on a particular issue. Keep redirecting your thoughts towards your problem or challenge. For some people it is helpful to note down the findings because it keeps them concentrated.

2.8.2. Problem finding

In addition, you need a scope (what do I want to achieve?), and a concrete problem or challenge (what exactly do I need to do or solve to achieve my objective?). Of course, you need information to be able to do this, but let us suppose you have done that already (you have already carried out a SWOT analysis for your company, or you have conducted interviews or desk research related to your scope). Then you need to find a particular problem, i.e. experiment with various ways of defining an issue to see what insights you gain.

This is easy to say, but practice shows us that this step is the hardest. 'The formulation of a problem,' said Albert Einstein, 'is often more essential than its solution, which may be merely a matter of mathematical or experimental skill. To raise new questions, new possibilities, to regard old questions from a new angle, require creative imagination and marks real advances in science.' Sharpen your focus and identify your real problem. In group situations, the definitions of the individual group members usually turn out to be different from each other, and these re-definitions

---

Blueprint for problem finding

Take a step back from your problem.

Look with an open mind.

Identify and break existing thinking patterns.

Think broadly.

Avoid generalisations (challenge yourself to be precise, ask yourself what things mean).

- Make a mind map or association field. Jot down issues related to your problem. Follow your thoughts when writing. Start out on a new line of thought when another one stops. Make connections where you see them. Establish an overview.
- Ask yourself 'Why?' three times in a row ('Why do you want to solve this question?' Answer A. 'Why A?' Answer B. 'Why B?' and so on).
- What is not the problem?
- What has been tried before? Why didn't it work?
- Picture the situation (i.e. explain it without words)
- Write down essential questions, starting with 'How do we...?' or 'How might we…?'
- Avoid the impulse to answer the question and see how many of these questions you can come up with. The more concrete the question, the more concrete your answer will be.
- Finally, choose the most appropriate one, the one that is really worth solving.

Bear in mind that this is an essential step. If you ask yourself 'How do I decide which is the best option?' then your answers will present you with the procedure for getting there, not the decision itself!

You might proceed by applying some techniques that are also used for problem solving.

Pinpoint your assumptions/prejudices (list those related to the problem) and then explore what happens as you drop each of these assumptions, individually or in combination.

Try defining what is unchangeable. Is it really unchangeable? Why?

Invert the situation ('How do we keep customers?' >> 'How do we scare them away?'). This will definitely present you with a fresh outlook on your problem.

---

therefore usually work as eye-openers for the 'problem-owner'. This illustrates the importance of communication, as mentioned earlier on. By the way, this part of the process is also really effective in any other 'ordinary' meeting. Establishing an agenda, not only with themes to be discussed but also what you expect from it (decisions to be made or options to be generated), and determining the 'real' problem before trying to solve it can prove to be a considerable time-saver.

2.8.3. Problem-solving / Spontaneous solutions

Start solving the problem. If you have performed the initial steps adequately, then you should already have some ideas. When stating the question, we make use of an automatism of the brain (on hearing a question the brain starts responding to it).

Then look for more alternatives. After a while you will come up with different ideas, and you will probably also find yourself examining some of the trains of thought raised in the former phase. If this flow of ideas comes to an end, or if you have no further thoughts or solutions to your subject, it is time to start using some creative thinking techniques.

2.8.4. Problem-solving / Creative thinking techniques

All techniques have a purpose. Creative thinking is basically about breaking through thinking patterns, and it is about making connections between contexts that were separate, until you stopped and looked again. This latter part is all about association and analogy. There are hundreds of techniques available, but they will not work unless you have the right question to work with. We will mention some of the most essential techniques. (If you are interested in more, please look at the reference list, or visit the websites mentioned). Allow your ideas to flow freely, without too much evaluation (this cannot be done, because...) – just for the moment.

Kinds of techniques

There are rational techniques, used to order your thoughts, identify factors and raise consciousness. They are logical, understandable and highly recommended for 'new' problems, and for people who do not particularly like 'flapping about'. Obvious or logical analogies are an example of these techniques. They can still be very imaginative: think about all the inventions we have made based on what we find in flora and fauna (Velcro, landing gear, etc.).

> *As an example of the use of analogy, Rice (1984) notes that the idea for Pringles potato chips came about through a search for an analogy to solve the problem of broken chips. Someone asked: 'What naturally occurring object is similar to a potato chip?' One suggestion was dried leaves. Exploration of this analogy led to a discussion about pressed leaves in a scrapbook or collection. This, in turn, led to the insight that the leaves must be pressed flat for storage whilst they are still moist;*

leaves will crumble if you wait until they are already dried. This idea became the basis for Pringles' innovative manufacturing and packaging process for its potato chips (Plsek, 1997).

Irrational techniques are meant to open up associative fields, just to inspire you. Free the bees for cross-fertilization. They are very effective for butterfly thinkers, and for illustrating the connection-making power of your brain. They can also be used for problems that have not yet been solved, even after long thought. Due to the 'mental leap', being forced out of the current situation, significant new insights can be gained simply by approaching the subject from a different angle.

Intuitive techniques are used to find what is hidden in your head, to divert your analytical brain. Copying a drawing or painting is effective, as is daydreaming. It will allow your brain to soften and let ideas arise 'naturally'.

Finally, there are techniques that are applied simply to stimulate a free flow of ideas and to create a comfort zone (intermezzo). They are used to stir your imagination (e.g. wishful thinking, or coming up with criminal or forbidden options or solutions) and to get your train of thoughts going.

Blueprint techniques (don't forget your preparation and evaluation!)
Change perspective
Reformulate your question into the opposite.
e.g.: How do we maintain a client relationship? >> How do we destroy it?
Find solutions for the new challenge.
Translate results back into original context.
N.B. Focus on one solution, then on another, think broadly, avoid generalisations.
Analogy
Choose a keyword or phrase in your challenge.
Look for similar situations, but in other contexts.
Find your inspiration in the world of technology, animals or humans.
Reformulate your question in the new context.
Find solutions for the new challenge.
Translate results back into original context.
Random association
Is your challenge clear and focused?
Bring in a random word (e.g. soup, sand, soap), or an object, photograph, smell, etc.
Associate freely as individuals.
Force connections to the problem.
Try explaining your ideas to the group or a 'buddy', as a soundboard.

2.8.5. Evaluation

Make an intuitive selection (ideas which seem attractive to you, whether feasible or not). Give unfeasible ideas a second chance before discarding them. 'Before you kill an idea, any idea, let's think of three reasons why it CAN be done (Michalko, 1991).' A helpful tool is to separate good from bad elements, by performing a P/M: add up all the plusses of the idea (why it would be good to do it, despite the problems we see connected to the idea), then add up all the minuses. Try to resolve the minuses. If the idea remains unfeasible, try recycling the good elements and use them to build another idea that is feasible.

Once you have determined your 'anchor points' (good ideas as the point of departure), the next step is to cluster other ideas or insights raised during the (individual or group) brainstorm around it, which should give you enough building blocks to start building a wall. You can only determine the potential of loose ideas after developing them into valuable concepts. Nevertheless, you will need to carry out a subjective evaluation before objectifying results. You will need to roughly determine whether the concept matches your strategies, whether it offers a real benefit to your customers and your company and whether it seems risky or feasible, before you spend resources on determining their real value with market research or test markets.

2.8.6. Concept development

Essentially, concept development in an early stage consists of making use of the generative mindset of the brainstorm. Try to find answers to who, what, when, where, why and how. At this stage it is also recommended that you focus on arguments to convince others of the idea's potential. This is necessary to make the idea a little stronger before 'throwing it out into the real world'. People who have not been present at the brainstorm session may not have enough essential information to understand the potential of an idea. Make sure they are given all the information they need to understand its value, but no more (keep it concise). When you confront 'outsiders' with ideas or concepts never ask: Shall we do this? Instead, ask: How could we improve on it? How could we make this happen? This way you sidestep everyone's instinctive habit of criticising and discarding before complimenting, and you rally the support for the project that you will need at a later stage.

In concept development for food innovation we focus on three areas, namely the marketing concept, the sensorial concept and the technical concept.
- The marketing concept
  The benefit to the consumer, the promise you make:
  - What problem does the product solve?
  - What need does it fulfil?
  - Who would be especially sensitive to this promise?

- The sensorial concept
  How can the consumer see/know that you are fulfilling the promise? Try to visualise the product itself, its packaging, name, etc. It does not have to be final, just give it a try and sketch the outline of possible ways of fulfilling the need or solving the problem.
- The technical concept
  Properties you need to make the promise come true.

Anchor points/ideas from the brainstorm can basically come from any one of the three fields. Make sure you fill in the blanks. You do not have to be a marketing specialist or a product developer in order to fill in specific fields. The key is that everybody tries to help each other and defines the outline of the concept. At this stage everybody is equal; in fact, everybody is a (potential) consumer of the product.

A good example of a concept that usefully illustrates these three aspects is 'Cornetto Soft'. Unilever faced stiff competition from fresh, soft ice cream for its pre-packed Cornetto product. Of course they were quite capable of making (ingredients for) fresh, soft ice cream, but one major obstacle was their brand. They could not afford any blemish on their reputation for quality ice cream, and soft ice is renowned for causing (temporary) illness, due to unsuitable hygiene conditions. It would not be feasible to control all points of sale, so they started looking for an alternative. They came up with a solution as follows:

- Marketing concept/benefit/reason to buy: fresh soft ice cream, safe, with the same taste everywhere (like the McDonalds concept), available everywhere (even at the smallest outlet).
- Technical concept: high quality, perfectly good soft ice cream, available in different flavours, pre-packed individually, like in retail outlets, BUT with one major difference:
- Sensorial concept: with the appearance of FRESH soft ice, because of its delivery. The ice cream is pushed out of the individual packaging by mechanical force, served from a machine that looks like a huge orange squeezer. It is being pushed into a real fresh cornet (thus solving one of the existing problems: Cornetto cones which stay in the freezer too long loose their crispness). The cornet is covered in a sleeve, because Unilever obviously still wants to communicate that this ice cream is from the Cornetto brand.

Summary of the idea generation process
Preparation/focus
1. Clarify your scope/briefing
2. Look at constituent parts
3. Reformulate (focus)
4. Choose one formulation of the problem
Idea generation
1. Focus and start solving (spontaneous solutions)
2. Use creative thinking techniques
a. Take a step back/change perspective
b. Break through your habitual thinking patterns (assumptions)
c. Find inspiration (look around you, find analogies)
d. Force associations
• Allow your ideas to flow freely
• Always pick up and write down ideas without judging them
• Always look for or 'force' a connection back to the problem
Evaluation
Make an intuitive selection (ideas which seem attractive to you, feasible or not)
Give unfeasible ideas a second chance before discarding them
These will be your anchor points
Group ideas into meaningful clusters
Concept Development
Bring ideas to life. Explore possible routes to the top.

2.9. In practice

Suppose you are responsible for product development, but you don't know where to start. There is a strategy but no ongoing projects. There might be plenty of ideas, but no concept. There are various technological possibilities, but no specific application. In such a situation, simply 'stating the problem' will not do you much good. You might end up asking yourself: 'How do I formulate a good concept?' But this question alone is not likely to give you the answer you were looking for: the concept itself. You know the principles (distance, focus, solve, etc.), but how do you establish that distance and how do you focus? Try this.

N.B. This is a good way to get started, but it is not the best way to run an innovation project. An innovation project starts with thorough preparation, making strategic choices, involving the right people, rallying management support, etc.

2.9.1. How to establish a distance?

What if I were... Imagine that you have been hired by an investor, a newcomer, to come up with a business plan to conquer your market within one year, rather than working within your current company. This will temporarily eradicate stumbling blocks such as existing strategies and production constraints.

What if 'we' had solved that particular problem, or if 'this' wasn't so. Make statements, as follows: Remember, back in 2002 (or in the year you are reading this text), when all we did was…, all we had was…, all we were doing… It will stir your imagination.

2.9.2. How to focus?

First, set out on a journey with adventure in mind. Open up to signals, e.g. think about consumer trends and developments in the market place. If you find this difficult, think of a product that is really new (introduced no more than one year ago) and attractive to you (a car, a watch, a mobile phone, a detergent, a food product, etc.). Which trends can you link to this product? Why did the producer introduce it and why is it successful? List the 'trends', and then choose one.

Or run through some newspapers or magazines, looking for clues. Anything that leads you to conclusions about the marketplace or inspires you to predict future developments, e.g. 'If they can do this, then their next step might be….' or 'If they want this now, we will soon have that too.' Make a list of statements, and then choose one. And at the same time, if you receive or pick up information from trend watchers, books or any other source, then don't just listen. Don't get buried under the heap of information. Instead, focus on one of the items and decide that you are going to translate that trend, or an interesting element, so that it applies to your product or market.

2.9.3. How to use this information from trend to product?

After you choose one trend or element (e.g. we need to make our pre-packed product fresher), you may want to start finding solutions, or you might take a step back and think a little further. Do you want to make a product which is fresher, or one which looks fresher? A product which can be preserved for a longer, or shorter time? A more natural product (ripened on the tree, rather than in a warehouse)? The perspective you adopt will influence the outcome. It will lead to more options, especially when you are stuck: 'This product cannot be made 'fresher'.' Or 'My product is not supposed to be fresh,' as, for example, with chewing gum. What if you were to manage to do it anyway? What would you do? Which other advantages and benefits would it give to your product? This is 'mental play'. Bear in mind that you are not evaluating as yet, you are not deciding what you are going to do. This is just opportunity scouting, and you are just roughly scanning possibilities.

Another technique is to look for examples in magazines, supermarkets and new product databases – in fact anywhere! Which other products are (perceived to be) fresh? Dairy products? Meat products? Fruit and vegetables? Fruit juices? Bread? Nuts? Why are they 'fresh'? As a consumer, how do we know they are 'fresh'? List all the reasons why, and then translate this to your product. You may want to offer your chewing gum from the fridge (opportunity for other spur-of-the-moment purchases, maybe even for ice cream), you may want to put it in transparent foil (opportunity to make nicer shapes and play with colours), you may want to make it softer to the touch, let it harden over time (so that consumers can determine how 'fresh' or 'mature' it is), or you might make it crispier or from naturally flexible ingredients, such as trees, bamboo, leaves, etc.

2.9.4. How to choose a direction?

Of course, if, after carrying out market research or analysis, you determine that a specific target group might need something special, you would direct your thoughts towards that target group. If you determine that there is a market out there that you had not noticed before, then you use your options to conquer that market. In the end it is all about concepts. Product, market and target group must be a perfect match! It doesn't really matter where you come from, or where you start. Objectives may surface after going on information-gathering adventures to the stores, or they may have been established beforehand. But in the end you must always apply your thoughts to achieving an objective, to scoring a goal. You need direction; otherwise you are likely to fire lots of blanks and ill-guided missiles! And you will end up being very busy, but without any sustainable results to show for it.

> *The myth goes that creatives either lie back and let the muse come to them, or force it out through hard work and lengthy trial and error. The reality is somewhere between the two – a combination of inspiration and evaluation – of being able to let an idea come to you and then crafting it into shape (Guy Claxton, Sunday September 22, 2002, The Observer).*

2.9.5. How to solve a specific problem?

Call to mind a particular project.
Define your key question.
Redefine your question in another context, to double-check the pertinence of your question and as inspiration for possible solutions. (N.B. Don't stop when you find one, but carry on and find another!)

Example: How might we reduce the percentage of waste on this line? You could reformulate this as follows: 'How might we reduce dog droppings on the street?' Or 'How do animals keep themselves clean?' (Reconsider: Do we want to keep the place clean, or do we want to reduce costs? If so, then do we achieve this by changing the ingredients or through cleaning activities?) We could give the animals less food, train them to go to a fixed spot, or use bacteria to harden droppings on the spot – so that people don't get messy shoes. This may lead you to the solution of starting off with fewer ingredients in your machinery, or a stickier product that doesn't disintegrate, installing a drain at a fixed spot or finding a way to render waste harmless.

As inspiration, as a way to flex your mind, think of how others have solved their problems, think of inventions or natural evolution in the animal kingdom. Make a list of what happened and why, and try to adapt it, and use it for your project. Make a list of several options, and then choose the ones you like best, or the ones you would like to start experimenting with.

These days, there are ongoing efforts to improve on innovation processes, and especially the soft part, the part where human 'messy' brains come in. Innovation, unlike audits or reengineering, is not given to formulas. But it helps to steer your brains into accurate directions.

2.10. How to hit the bull's eye

Systematic Inventive Thinking (SIT) is a controlled brainstorming process. SIT is an efficient way to innovation as it delivers quick, strong, and commercially interesting results.

Over the last few years, some developments have taken place. As a response to our productivity-oriented society, the University of Tel Aviv has worked on a methodology to make brainstorming more effective. As a result, a particular thinking process has been developed, which deserves attention. We have come to call this process 'second generation brainstorming' as it has nothing to do with the original 'wild' brainstorming practices of the first generation. On the contrary, it is a highly controlled process. SIT takes a new perspective on concept- and product development. Over the last decades, the focus has been on consumer trends and consumer insights as the ideal starting point for brainstorming sessions. SIT takes the existing product itself and its route to market as a starting point. It is not that consumer insights and trends are being ignored, they are just not the starting point for the thinking process.

Resource-oriented methodologies: from a product and process point of view: What are the crucial components (resources) in my product and process? What else can I do with it? Which parameters can I change, with a positive (and without a negative) impact on my product and brand?

Market-oriented methodologies: from a consumer point of view: What is happening around me, in this world? What can I do to translate this to my product/brand, based on my core competences and resources?

2.10.1. How does it work?

Instead of looking away from the product, into the wide world, the focus is on the product and its production process itself. The core of the process is to dissect products and processes into components. These components are then re-arranged, along predefined routings or so-called 'templates' (inventor's thinking patterns). Then the team starts looking for advantages in the new product or process. Advantages can be found in price (cost-saving measures) or in benefits (value-adding measures). If no advantages are found, the idea gets discarded immediately, and the team proceeds to a new assignment with another component or another template. If, on the other hand, an advantage is found, the idea is developed into an attractive and feasible concept.

These 'templates' are based on thorough research into product evolution, as found in patents. There is a repeating structure here, as it seems there is a predefined path for new thoughts to develop in your mind. Using these patterns as a guideline, you will develop truly creative and innovative products, yet recognizable ones. Consumers will recognize the underlying patterns in the thought process which led to the product idea. Thus, the product will be regarded as new and surprisingly well thought of, meeting true consumer needs. And it will find immediate acceptance in the market, as it is a perfectly logical development of its original.

2.10.2. Why does it work?

People tend to look for solutions where they expect one. This means they do not think freely and objectively, they are caught in their existing thinking patterns. The SIT methodology makes people change their 'usual', intuitive thinking routines in a highly effective way, leading to ideas with a high degree of newness and surprise. A positive side-effect of this recently developed thinking technique is that, because of its analytical character and strict assignments (according to predefined patterns for change), every person, regardless of his or her natural thinking style, adopts the patterns very easily.

In the context of group brainstorming, this is of tremendous value. After all, the main essence of any (group) thinking technique is to create a shared frame of reference, a shared language, and a shared goal. The SIT process stimulates knowledge exchange which helps people to find original solutions to solve problems and seize opportunities. Because of the involvement of stakeholders in the process, the solutions are actually adopted and therefore swiftly implemented. This, together with the immediate acceptance in the market, is the essential ingredient for successful innovation in the 2000s.

Main templates (patterns of evolution in products and processes, Figure 2.1):
* subtraction: remove a component (e.g. instead of constantly adding functions to mobile phones, remove them in order to make a simple, easy-to-use mobile phone, for kids or elderly);
* multiplication: multiply a component and adapt (e.g. multiply razor blades, and reposition one of them to make the razor effective for side burns);
* division: rearrange components as if they were separate entities (e.g. separate adhesive components for strong effects when mixed);

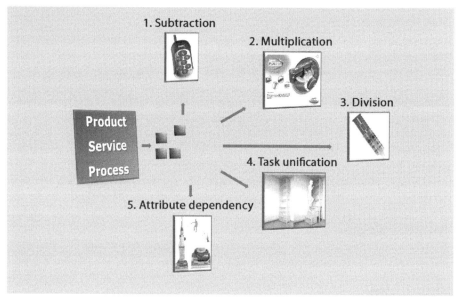

Figure 2.1. Patterns of evolution in products and processes.

- task unification: add and re-arrange functions of components (e.g. use drawers of a cupboard as steps for the hard-to-reach top cabinets);
- attribute dependency: play with variables of components (e.g. adopt length of vacuum cleaner's hose for an easy stair vacuum).

References

Michalko, M. (1991). Thinkertoys. Berkeley.

Osborn, A.F. (1993). Applied Imagination. Creative education foundation press, Buffalo, New York. Third revised edition.

Parnes, S. (1997). Optimize the magic of your mind. Bearly Ltd, Buffalo, New York.

Plsek, P.E. (1997). Creativity, Innovation, and Quality. ASQ Quality Press.

Wycoff, J. (1991). Mindmapping (originally by Tony Buzan). Berkley Books, New York.

Further information

Amabile, T. (1996). Creativity in Context. Westview Press, Inc.

Edwards, B. (1987). Drawing on the Artist Within. Simon & Schuster, New York

Gelb, M.J. (2000). Discover Your Genius: How to Think Like History's Ten Most Revolutionary Minds. Dell Books.

Higgins, J. (1994). 101 Creative Problem Solving Techniques: The Handbook of New Ideas for Business. New Management Pub. Co.

Higgins, J. (1995) Innovate or Evaporate. Test & Improve Your Organization's IQ: Its Innovation Quotient. New Management Pub. Co.

Johansson, F. (2004). The Medici Effect: Breakthrough Insights at the Intersection of Ideas, Concepts, and Cultures. Harvard Business Press.

Kanter, R.M. (1984). The Change Masters: Innovation and Entrepreneurship in the American Corporation. Simon & Schuster, New York.

Koestler, A.(1982). The Act of Creation. First published by Hutchinson 1964, Penguin.

Morgan, G. (1993). Imagin-i-zation: New mindsets for seeing, organizing and managing. Sage Publications, California.

Zakia, R.D. (2001). Perception & Imaging. Butterworth-Heinemann (Trd). 2nd edition.

You can reach the author at corrinne@whitetree.com.

Websites

www.directedcreativity.com
www.gocreate.com
www.sitsite.com
www.whitetree.com

3. Structured food product development based on Quality Function Deployment

Marco Benner, Matthijs Dekker and Anita Linnemann
Product Design and Quality Management Group, Wageningen University

3.1. Introduction

The first chapter of this book explains the need for continuous new food product development, and in the second chapter we describe the initial step of new food product development, namely the idea generation process. This chapter presents a tool that can be used to develop a product concept into a set of tangible product characteristics, called Quality Function Deployment (QFD). QFD is an approach which ensures that customer or market requirements are systematically translated into relevant technical requirements and actions through the consecutive stages of product development. As such, it is a planning method that guarantees that quality is engineered into a product at the design stage (Charteris, 1993). QFD is an adaptation of tools used in Total Quality Management (TQM). The method relates different types of data, namely consumer information and technological know-how, to encourage product development team members to communicate more effectively with each other. It helps teams to formulate business problems and their solutions (Cohen, 1995). The ultimate goal of QFD is to assist companies in producing exactly the new product that the consumer requires and in reducing the time-to-market. In other words, the method improves the profitability of companies by increasing the success rate of product development.

QFD was initiated in the late nineteen-sixties in Japan to support the product design process. At first it was used for the design of large ships at Mitsubishi's Kobe shipyard. As it evolved, it became clear that it could also be used to support service development. QFD has been extended to apply to any planning process where a team wants to systematically prioritise their possible solutions to a given set of objectives (Urban and Hauser, 1993). QFD has been widely spread among industries in the western world since its introduction in the USA in the beginning of the nineteen-eighties. Among the first users of QFD were companies like Ford Motor Company, Procter and Gamble, Campbell's, IBM, Xerox, Hewlett-Packard, Kodak, and 3M Corporation (Griffin and Hauser, 1993; Cohen, 1995).

This chapter describes the QFD approach, how it can be used, what its major benefits and drawbacks are, and its future perspectives. The chapter is partly based on previous publications by the authors on QFD, namely with respect to the stepwise application of the method (Dekker and Linnemann, 1998), the specific characteristics of the approach for the food industry (Benner *et al.*, 2003b) and research into improvements on the QFD method (Benner *et al.*, 2003a).

3.2. Quality Function Deployment step by step

3.2.1. House of Quality

The QFD method begins with the construction of the so-called Product Planning Matrix, often referred to as the 'House of Quality' because of its contents and the fact that its appearance resembles the sketch of a house. The House of Quality consists of several so-called rooms, each containing information concerning the product that is to be developed. The main goal is to translate the customer demands into specified product requirements. The basic structure of the House of Quality is presented in Figure 3.1.

3.2.2. The voice of the customer (WHATs)

The first room of the House of Quality consists of a set of decisive customer requirements, briefly described as the WHATs. These requirements are also known as 'the Voice of the Customer' or the desired quality characteristics of the product. The information on customer demands is usually obtained by market research, focus group interviews and other means to assess customer priorities. The WHATs are often, at least initially, described in the customers' own words, and may therefore be vague and general. Customer requirements need to be clearly defined before they are inserted in the House of Quality. Other customer wants are not verbalised, but are obtained by observation (Hofmeister, 1991). Complaints about previous similar

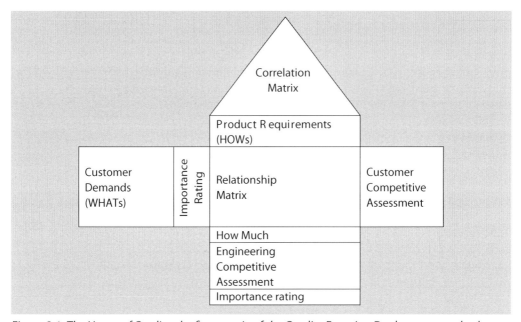

Figure 3.1. The House of Quality, the first matrix of the Quality Function Deployment method

products also are a valuable source of information. The customer demands are rated against each other to quantify their importance in realising the ultimate success of the product. This importance rating, frequently expressed on a scale from 1 to 5, can help to set priorities for the product development process and provide guidelines for allocating the necessary resources. The data on the importance rating are inserted in the House of Quality in a room adjacent to the room with the customer demands.

3.2.3. Product requirements (HOWs)

Once the relevant customer requirements are identified, each of these must be elaborated to explain what a particular demand will imply for the technological characteristics of the product itself: WHATs must be translated into product requirements, also called the HOWs. The HOWs should be measurable properties that describe the desired customer want in the technical language of the company. A HOW should be read as: 'how to measure' and not 'how to accomplish'. The product requirements may not be a set of ingredients or process parameters, and therefore still leave scope for creativity concerning aspects related to design and manufacture of the product (Hofmeister, 1991). An example of a translation of a customer demand into a product requirement can be derived from consumer's wishes with respect to convenience food. A customer will ask for 'a short preparation time', which leads to a product requirement of, for instance, a 'maximum of 7 minutes at 700 W' to heat the product in a microwave oven.

3.2.4. Relationship matrix

The centre part of the House of Quality contains the relationships between the WHATs and the HOWs section. This so-called relationship matrix provides a crosscheck, since it may not contain empty rows or columns. An empty row indicates that the company has omitted to specify a product requirement that is necessary to meet a customer demand regarding the product. An empty column implies that one of the technical product requirements does not contribute to the customer demands with respect to the product. In the relationship matrix symbols can be used to indicate the strengths of the relationships between a particular customer demand and a product requirement. Usually four levels of strengths are applied, namely strong (++), medium (+), weak (+-) and none (0). This rating can again assist in setting priorities for the product development process and in the allocation of financial and human resources.

3.2.5. Correlation matrix (the roof of the house)

On top of the House of Quality is a triangular table, which is called the correlation matrix. This roof-like table shows the correlations that exist between the different product requirements. An alteration in a particular product requirement that is related to another product requirement will usually affect both requirements. Correlations identified in the roof indicate areas where trade-off decisions and research and development are needed. Here too, symbols can be used to indicate the nature and the strength of the correlations. Usually four denotations are used, namely for a) a strong positive correlation (++), b) a weak positive correlation (+), c) a weak negative correlation (-), and d) a strong negative correlation (--). Strong positive correlations imply that an improvement in one product requirement supports another. This information is important. It offers the opportunity to use research funds efficiently by avoiding duplication of efforts to achieve the same result. Strong negative correlations demand attention because they represent conditions in which trade-off decisions may be needed. In such a case it is necessary to determine whether both target values can be achieved or whether more innovative research can assist in passing technological barriers. As such, the correlation matrix is valuable for identifying long-term research objectives that are linked to customer demands.

3.2.6. Objective target values (HOW MUCHs)

The 'cellar' of the House of Quality contains several rooms with different types of information. One section contains the HOW MUCHs. Here, target values are given for all product requirements (HOWs). The HOW MUCHs should explicitly represent the customer demands and certainly not the current performance levels of the product. The HOW MUCH section of the House of Quality provides univocal objectives to drive the subsequent product development by means of objectively setting goals to monitor progress.

3.2.7. Competitive assessment

It is important to know how competitive products compare with current company products in order to (1) establish proper target values (HOW MUCHs) and (2) ensure a good correlation between the WHATs and the HOWs. This comparison is carried out for the WHATs and the HOWs separately. The competitive assessment of the WHATs is called the customer competitive assessment. This information must be customer-oriented and gathered through market surveys and similar means. The customer ratings should be collected for the company's product and the competitive products on each of the customer demands (WHATs), for instance on a scale from 1 (worst) to 5 (best). Analytical methods to relate the customer ratings to the engineering properties of the products should be the company's own techniques. The values for the customer competitive assessment are filled in on the right-hand side of the House of Quality (Figure 3.1).

The competitive assessment of the HOWs is called the engineering competitive assessment. The competitive products are analysed using the company's own techniques to obtain this information. Hofmeister (1991) strongly recommends that the product engineers are directly involved in this process in order to gain the most complete understanding of the competitive products. This information is put in the 'cellar' of the House of Quality, below the HOW MUCHs (Figure 3.1).

The two sets of data, i.e. on the customer competitive assessment and the engineering competitive assessment, yield very important information. In case a WHAT and a HOW item are strongly related to each other, then the customer and engineering competitive assessments should be consistent with each other. If a combination of a WHAT and a HOW has more or less opposite scores, for instance, a 'worst' score in the customer competitive assessment for a WHAT and a 'best' score in the engineering competitive assessment for the related HOW, something is seriously wrong. Before continuing with the product development process, this problem of inconsistency must be investigated and sorted out. If such a conflict is not solved properly, then there is a serious risk that the company will develop the best possible product according to in-house tests, but not according to customers' wishes.

3.2.8. Setting priorities

A complete House of Quality contains two sets of importance ratings. The first has already been mentioned in the paragraph on the Voice of the Customer (WHATs); it signifies the relative importance of the different customer demands in realising the success of the product, usually expressed on a scale from 1 to 5. The values are obtained by market research.

The second importance rating is calculated using the customer importance rating and data from the relationship matrix in the centre of the House of Quality. This importance rating is called the technical importance rating. First the relations in the matrix are given a numerical value, e.g. a strong relation (++) gets 9, a medium strong relation (+) 3 and a weak relation (+-) gets 1. Technical importance ratings are obtained by multiplying the value of the customer importance rating with the value for the strength of the relationship for all combinations of a WHAT and a HOW in the matrix. For each HOW column a total value is calculated by summing up, and this total is filled in on the bottom level of the 'cellar' of the House of Quality. These totals are used as a guideline for setting priorities for product development (Hofmeister, 1991).

3.2.9. Cascading to the next 'Houses'

After the House of Quality is constructed, other matrices for design, process, production and packaging of the product should be built to complete the QFD approach. This cascading from one matrix to the next is done by using the HOW (MUCHs) of a chart as the WHATs of the following matrix (Figure 3.2). Items that are used as input for a following matrix, should be selected carefully to keep the cascading process manageable. Only new, important and difficult to achieve HOWs are transferred to the next matrix. Usually two different development routes are derived from the House of Quality, one for the product and one for the package (Hofmeister, 1991). In the food development route the House of Quality is consecutively followed by matrices on ingredients and process design and on production design and planning. The package development route usually consists

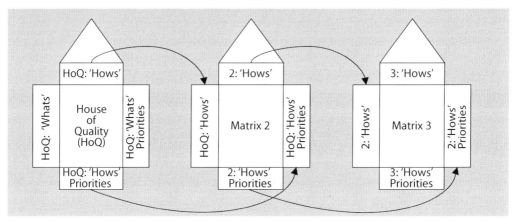

Figure 3.2. Cascading from the House of Quality to the next matrices in the application of the Quality Function Deployment method.

of the following matrices: product planning (i.e. House of Quality), package design, manufacturing process design and production design and planning.

3.3. Benefits and limitations of the Quality Function Deployment method

The application of the first matrix of the QFD method, i.e. the Product Planning Matrix or the 'House of Quality', reduces design time and costs in food product development. This financial advantage is attributed to increased communication among company divisions early in the product development process and the assurance that the voice of the customer is built into the development process (Urban and Hauser, 1993; Hauser and Clausing, 1988). The major benefits of using the House of Quality (Sullivan, 1986; Hauser and Clausing, 1988; Griffin, 1992; Hauser, 1993; Govers, 1996) are that:

- The House of Quality increases customer satisfaction by making sure that 'the voice of the customer' is systematically used to drive all aspects of the product development process.
- The House of Quality brings together all the data required for the development of a good product.
- The information input has to be concise and well-defined, and therefore is generally well thought-out and substantiated.
- The development team sees quickly and easily where additional information is needed for the development process.
- The application of the House of Quality improves effective communication between company divisions (e.g. marketing and R&D) and enhances team work.
- Important production control points are not overlooked.
- Company teams are working on avoiding problems instead of having to solve them later on.

- The House of Quality helps companies to make the key trade-offs between what the customer demands and what the company can afford to produce.
- Quality is built in upstream.
- The application of the House of Quality shortens time-to-market.

The construction of the House of Quality is the first step of the QFD method. Then the other matrices for design, process, production and packaging of the product should be built. However, in our experience this is too complex and therefore time-consuming for food products. We attribute this difficulty to specific differences between industrial goods and food products (Benner *et al.*, 2003), namely:

- food ingredients show a natural variation in composition; one batch of raw materials may differ significantly from the next batch of the same ingredient;
- many food ingredients are still physiologically active, which leads to quality changes in the course of time;
- interactions take place between the ingredients during the production process;
- processing influences the properties of food products.

3.4. The Chain Information Model

3.4.1. The information matrix

The Chain Information Model (CIM) was developed on the basis of the House of Quality. The CIM is a structured approach for gathering and disseminating information that is essential for an effective and efficient food product improvement process in food production chains. The CIM provides the information needed by each actor in the production chain for the successful development of a product, structures this information and filters out the important pieces. The CIM achieves this goal by subsequently seeking answers to the following research questions:

- What are the quality characteristics of a successfully improved product?
- How do the actors in the production chain contribute to this improvement?
- Which scenarios are available for realizing the intended product?
- Which scenario is the best?
- Who needs what information from whom to obtain the best scenario?

First, the so-called Information Matrix is constructed by replacing the product requirements in the House of Quality of the QFD method by the actors of the production chain (Figure 3.3).

Thus, a direct link is created between the customer demands for the product (WHATs) and the chain actors (WHOs). In the central part of the matrix the relationships between the customer demands and the actors are indicated. Furthermore, the central part of the matrix can be used to indicate the strength of the relationships (or dependencies) between the actors and the customer demands.

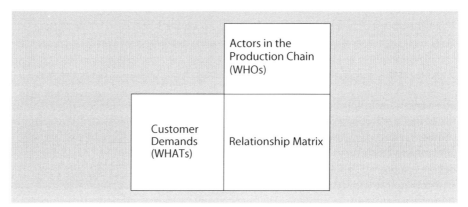

Figure 3.3. The information matrix.

3.4.2. Chain Information Model (CIM)

The CIM consists of three phases: (1) the information-gathering phase, (2) the information-processing phase, and (3) the information-dissemination phase (Figure 3.4). In the information-gathering phase (phase 1), the information needed for the effective development of an improved product is gathered. Three kinds of information are required to complete this phase. First, quality characteristics that make the current product successful are determined. Secondly, the current production chain, including all actors and production processes, is mapped out. Finally, information on the influences of the entire production chain on a desired new product feature is collected from written sources and by expert consultation.

In phase 2 of the CIM, the information-processing phase, all information from phase 1 is processed into essential information for the actors in the production chain. In this phase, the influences of the processes in the entire chain, and hence the influences of the actors, on the new product feature and the quality characteristics of the product are analysed. The influences are placed in a matrix that shows if and to what extent the actors influence the different quality characteristics.

In the information-dissemination phase, phase 3 of the CIM, the information matrix from phase 2 is used to select the actors with the largest potential influence on the desired new product feature. Next, scenarios are generated to develop the intended product. Each scenario describes options for each actor to produce the intended product. The scenarios also analyse and describe what the results are of changes in the production process on the other quality characteristics, and what the consequences are of these changes for the other actors in the production chain. The scenarios also indicate how processes should be changed to optimise the production process for the intended product and show which information should be shared for its realization (Benner *et al.*, 2003).

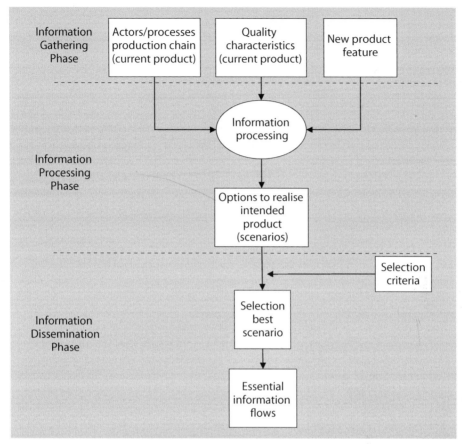

Figure 3.4. The Chain Information Model (CIM).

3.4.3. The information-gathering phase

Figure 3.5 presents an example of the possible outcome of the information-gathering phase of the CIM. This Information Matrix shows the required information to produce a tomato ketchup with a health benefit to its consumers in the form of an increased amount of lycopene. A distinction is made between actors who have influence on the quality characteristics (indicated with a '+') and actors that do not have an influence ('-') on the quality characteristics. In Figure 3.5 several actors are depicted that might have an influence on the quality characteristics but cannot actively influence them; these actors are supposed to follow the procedures that come with the product, for instance, on handling and temperature.

The information in the matrix concerning the consumer demands is assessed in the same way as for the construction of the House of Quality in the QFD approach. Information on the production chain and on the processes in the production chain that influence the quality characteristics, is collected from written sources and by

Quality Characteristics		Breeder	Grower	Transport[*]	Paste producer	Transport/ storage	Ketchup producer	Transport/ storage[*]	Retailer[*]
Easy to serve	Flows easily from the bottle	+	+	-	+	-	+	-	-
	Pours without scattering	+	+	-	+	-	+	-	-
Healthy	Amount of lycopene	+	+	-	+	+	+	-	-
	Bio-availability lycopene	-	-	-	+	-	+	-	-
	Contains no preservatives	-	-	-	+	-	+	-	-
	No thickeners	+	+	-	+	-	+	-	-
	Contains no fat	-	-	-	-	-	+	-	-
Appearance	Natural colour	+	+	-	+	-	+	-	-
Tasty	It is thick in the mouth	+	+	-	+	-	+	-	-
No defects	Never spoils	-	-	-	+	-	+	-	-
Clear information	Proper storage instructions	-	-	-	-	-	+	-	-
Best package	Can see product inside	-	-	-	-	-	+	-	-

- : no influence
+ : influence
[*] : these actors can not influence the quality characteristics

Figure 3.5. The information matrix for tomato ketchup with a health benefit due to the presence of lycopene.

expert consultation. Subsequently, this information is used to compose so-called Quality Dependence Diagrams (QDD), which link a consumer demand to the actors in the chain that have an influence on that specific quality characteristic. Figure 3.6 provides an example in which the health benefit derived from the presence of lycopene in tomato ketchup is linked to the actors in the chain that may influence the amount of these beneficial components.

3.4.4. The information-processing phase

After all the required information is gathered, it has to be combined and evaluated to determine the possible scenarios for producing the intended product. These scenarios are formulated following a systematic analysis of the options for every actor in the production chain to realize the required amount of lycopene in the tomato ketchup. For every possible change made by an actor, the consequences for the other quality characteristics and for the other actors have to be identified. Decision trees were developed to systematize the assessment of all feasible scenarios. To clarify this

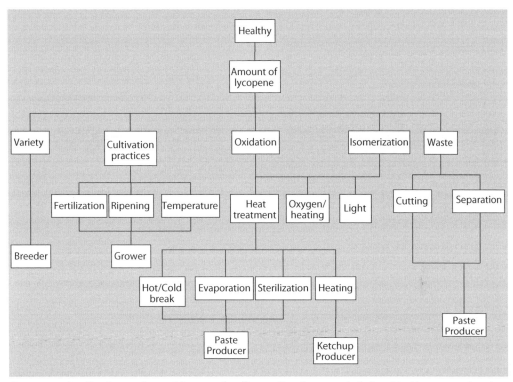

Figure 3.6. Quality Dependence Diagram for the quality characteristic 'healthy' in tomato ketchup with a health benefit due to the presence of lycopene.

method, the possibilities are elaborated for the first actor in the production chain of the tomato ketchup with a health benefit, namely the breeder of the tomato.

The breeder can develop, or select if available, a variety with a higher lycopene content. The consequences of this scenario are further elaborated in the decision tree in Figure 3.7. Selection of a different variety with higher lycopene content is the easiest way, but will result in other altered characteristics of the tomato. With conventional breeding programs it is possible to raise the amount of lycopene without changing other characteristics of the tomato. The major drawback is that it takes several years before a new variety is developed. The third option is genetic modification; this method takes less time than conventional breeding but has a low acceptability among consumers (Schifferstein *et al.*, 2001).

3.4.5. The information-dissemination phase

The decision trees have to be constructed for all the actors that can influence the quality characteristic 'amount of lycopene'. This results in several scenarios for the production chain to raise the amount of lycopene. In order to determine the best

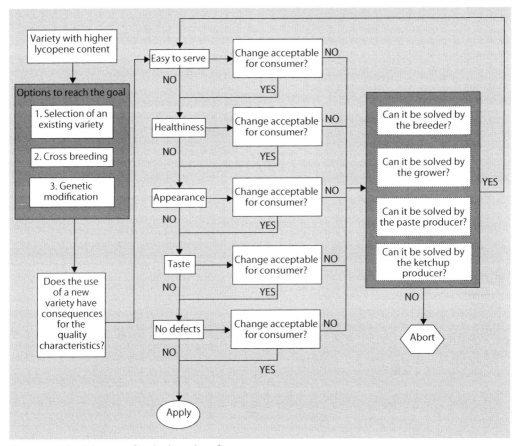

Figure 3.7. Decision tree for the breeder of tomatoes.

scenario, the alternatives have to be prioritized against criteria that are defined by the actors in the production chain. Profitability and technological feasibility are important criteria when selecting the most preferable scenario. The selection criteria depend on the role of the actors in the production chain. An optimal situation for the production chain as a whole does not necessarily implicate an optimal situation for all individual actors. Producing a successful product might imply that some actors have to invest more and get less for their money in the short term. For the actors involved in production a common long-term strategy has to be developed. Once the best scenario has been determined it has to be incorporated into the production chain by informing each actor accordingly.

References

Benner, M., Geerts, R.F.R., Linnemann, A.R., Jongen, W.M.F., Folstar, P. and H.J. Cnossen (2003a). A chain information model for structured knowledge management: towards an information supply model for effective and efficient food product improvement. Trends in Food Science & Technology, 14: 469-477.

Benner, M., Linnemann, A.R., Jongen, W.M.F. and P. Folstar (2003b). Quality Function Deployment (QFD), can it be used to develop food products? Food Quality and Preference, 14: 327-339.

Charteris, W.P. (1993). Quality function deployment: a quality engineering technology for the food industry. Journal of the Society of Dairy Technology 46(1): 12-21.

Cohen, L. (1995). Quality function deployment: how to make QFD work for you. Addison-Wesley Publishing Company, Reading.

Dekker, M. and Linnemann, A.R. (1998). Product development in the food industry. In: W.M.F Jongen and M.T.G. Meulenberg (eds.) Innovation of food production systems: product quality and consumer acceptance. Wageningen Pers, Wageningen. pp. 67-86.

Govers, C.P.M. (1996). What and how about Quality Function Deployment (QFD). International Journal of Production Economics (46/47): 575-585.

Griffin, A. (1992). Evaluating QFD's use in US firms as a process for developing products. Journal of Product Innovation Management 9: 171-187.

Griffin, A. and Hauser, J.R. (1993). The Voice of the Customer. Marketing Science 12(1): 1-27.

Hauser, J.R. (1993). How Puritan-Bennet used the House of Quality. Sloan Management Review (spring): 61-70.

Hauser, J.R. and Clausing, D., (1988). The Voice of the Customer. Harvard Business Review 66(5/6): 63-73.

Hofmeister, K.R. (1991). Quality Function Deployment: market success through customer driven products. In: E. Graf and I.S. Saguy (eds.) Food product development: from concept to the market place. Van Nostrand Reinhold, New York. pp. 189-210.

Schifferstein, H.N.J, Frewer, L.J., and Risvik, E. (2001). Introduction. In: L. Frewer, E. Risvik, H. Schifferstein (eds.) Food, People and Society: A European perspective of consumers' food choices. Springer.

Sullivan, L.P. (1986). Quality Function Deployment: a system to assure that customer needs drive the product design and production process. Quality Progress (June): 39-50.

Urban, G.L. and Hauser, J.R. (1993). Design and marketing of new products. 2nd edition. Prentice Hall, Englewood Cliffs, N.J.

4. Modelling of reactions affecting food quality

Martinus A.J.S. van Boekel
Product Design and Quality Management Group, Wageningen University

4.1. Introduction

As indicated in Chapter 1, there are several reasons why product and process design needs to be flexible these days. Gone are the days when this could be done by trial and error. A more systematic way is by use of modelling; in fact one could think of the design process as being done in a virtual lab using a computer. The ultimate goal of food product design is to design a product (along with the associated processes) that will satisfy consumer wishes, which implies that it needs to be clear which quality attributes should be present in the food. Because of chain reversal, modelling of the whole food chain might even be necessary. When one tries to model this whole integrated process, the need for knowledge management becomes obvious, as discussed in Chapter 9. Modelling of all the processes in the chain, including consumer wishes is not yet possible, but there are new developments on the horizon, such as Bayesian Belief Networks and neural networks. This chapter will not discuss these new developments in great detail but will focus on the current possibilities to model product quality attributes. First, it is essential to know what to model. We start therefore with a short overview of key reactions that have an effect on quality. After that we give a general description of models and some possible applications in product design. It is essential to realize that models for product design are based on experimental work. Therefore, the translation of experimental work into useful models is a critical step. The reader is referred to a recent textbook on this aspect (Van Boekel, 2008).

4.2. Key reactions in foods

The stability and quality of foods can ultimately be related to chemical, biochemical, microbial and physical reactions. Many textbooks describe these reactions, for instance, Fennema (1996), Owusu-Apenten (2005), Damodaran *et al.* (2009) focus on chemical reactions, Whitaker *et al.* (2003) on biochemical reactions, Walstra (2003) on physical reactions, while Jay *et al.* (2005) is a reference for microbial reactions. Knowledge about these reactions is obviously of importance for food design. In fact, it is important to know which reactions to expect for a certain design. For instance, if one designs a food that contains reducing sugars and proteins and/or amino acids, the Maillard reaction is bound to occur. If a food is to contain unsaturated fatty acids, oxidation is going to be a problem. Raw materials of plant and animal origin will usually contain active enzymes causing changes. Two things are necessary in this respect:

1. Knowledge on the type of reactions and their consequences for quality.
2. Knowledge on how fast these reactions occur.

A study on change in quality is in fact a study in kinetics. Table 4.1 gives an overview of the most important reactions in foods.

Another way to look at reactivity is to look at the various ways in which the main components in foods can react: see Table 4.2.

Most foods contain water (except pure fats and oils). Human beings soon discovered that drying was a very effective means of food preservation. The role of water in food stability is recognized as very important. Every food scientist knows the 'food stability map': see Figure 4.1.

The dependence of microbial growth on water activity is well-known as a phenomenon, but it is not well understood. It probably has to do with osmotic effects. Still, many compounds that lower water activity also have other specific effects on micro-organisms and this is often species-specific, which makes it difficult to predict. Nevertheless, the food stability map is able to predict rates of microbial reactions

Table 4.1. Overview of key reactions in foods.

Example	Type	Consequences
Non enzymatic browning	Chemical reaction (Maillard reaction)	Colour, taste and flavour, nutritive value, formation of toxicologically suspect compounds (e.g. acrylamide)
Fat oxidation	Chemical reaction	Loss of essential fatty acids, rancid flavour, formation of toxicologically suspect compounds
Fat oxidation	Biochemical reaction (lipoxygenase)	Off-flavours, mainly due to aldehydes and ketones
Hydrolysis	Chemical reaction	Changes in flavour, vitamin content
Lipolysis	Biochemical reaction (lipase)	Formation of free fatty acids, rancid taste
Proteolysis	Biochemical reaction (proteases)	Formation of amino acids and peptides, bitter taste, flavour compounds
Enzymatic browning	Biochemical reaction of polyphenols	Browning
Separation	Physical reaction	Sedimentation, creaming
Gelation	Combination of chemical and physical reaction	Gel formation
Acidification	Microbiological (fermentation of sugars)	Formation of acids, flocculation of proteins at low pH, sour taste
Spoilage	Microbiological (growth of moulds, yeasts, bacteria)	Taste and texture changes, food poisoning, appearance

Table 4.2. Reactions of key components in foods.

Component	Reaction	Consequences
Proteins	Denaturation	Gelation, precipitation, solubility, inactivation of anti-nutritional factors (ANFs)
	Hydrolysis	Formation of peptides, amino acids
	Deamidation	Loss of charge and change in reactivity
	Maillard reaction	Crosslinking, loss of nutritional value, browning
Lipids	Oxidation	Loss of essential fatty acids, rancidity
	Fat hardening	Formation of trans fatty acids
	Hydrolysis (usually enzymatically)	Formation of free fatty acids, leading to a soapy off-flavour
Mono- and disaccharides	Maillard reaction	Non-enzymatic browning
	Caramelization	Taste and flavour changes
	Hydrolysis	Sugar inversion
Poly-saccharides	Hydrolysis (enzymatically during ripening, chemically during cooking)	Softening of tissue
	Physical interaction with other components	Gelation, phase separation
	Gelatinisation and retrogradation of starch	Staling of bread
Polyphenols	Enzymatic polymerization	Browning
	Interaction with proteins	Crosslinking, gelation
Vitamins	Oxidation	Loss of nutritional value

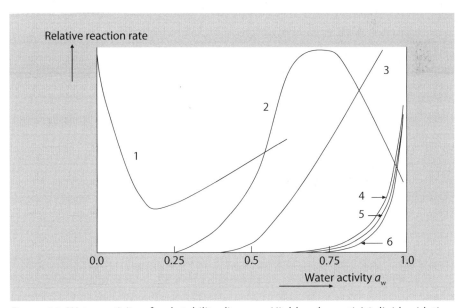

Figure 4.1. Water activity - food stability diagram. Highly schematic! 1: lipid oxidation. 2: non-enzymatic browning. 3: enzyme activity. 4: mould growth. 5: yeast growth. 6: bacterial growth. At a_w=1 (pure water), all rates become zero.

at least in a qualitative way. The dependence of enzyme activity on water activity as depicted in Figure 4.1 can be explained, qualitatively at least, because pH, ionic strength and solute activity coefficients change with water activity, and this can have great effects on protein conformation and hence on enzyme activity. Furthermore, water content can have an influence on the effective diffusion. At low water content, diffusion may be hindered considerably, thereby slowing down reaction rates. This is also the qualitative explanation for the dependence of the non-enzymatic browning reaction as shown in Figure 4.1. In going from a high to a low water content the reactants become more concentrated and the rate will increase up to a point where the reaction becomes diffusion limited: diffusion becomes increasingly difficult when the water content is further reduced and as a result the rate decreases. When water acts as a catalyst or an inhibiting agent, changes in water content will also have an effect on kinetics in that respect, as discussed above. A case in point is the effect of water on fat oxidation. Oxidation rates will increase with decreasing water content probably because water acts as an inhibitor for oxidation. All these effects interfere and it is therefore not possible to predict exactly what will happen. Trends may sometimes be predicted but it is very difficult to make predictions in a quantitative way for real foods.

However useful the food stability map is, it is more a phenomenological than a mechanistic way of looking at stability. Water activity is really a thermodynamic concept, and relates to equilibrium situations. This is hardly the case for foods. Water activity is in principle a measure for the chemical potential of water. A practical way to determine water activity a_w is by measuring the water vapour pressure above a food p_w and relating that to the saturated vapour pressure of water p_w^0:

$$a_w = \frac{p_w}{p_w^o} \tag{4.1}$$

In an ideal solution, water activity equals the mole fraction of water ($a_w = X_w$) in the solution, but in non-ideal solutions water activity cannot be predicted theoretically from the composition of a solution. Ideal solutions obey Raoult's law, but obviously foods cannot be considered ideal solutions. Even simple solutions show deviation from ideality: see Figure 4.2.

The example given in Figure 4.2. makes clear that nonideality can be very pronounced, and the big difference between the two, seemingly very identical, molecules glucose and fructose is striking. This should demonstrate that it really is impossible to predict water activity in a food with its many constituents; the only way left is to determine water activity by experiment. However, many empirical relations are available in literature to predict water activity in a particular food. These should be used with caution. Frequently, relationships are proposed between water activity and a certain reaction, but these are usually only correlations. On a mechanistic basis, water activity can have an effect on a chemical reaction if water takes part in the reaction. Otherwise, water activity is a factor that relates to another factor. For instance, with decreasing water activity, reaction rates usually decrease. One reason for this is that

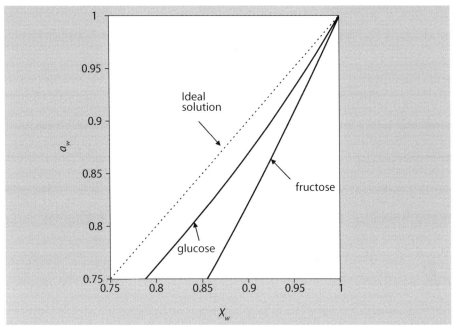

Figure 4.2. Examples of non-ideal behaviour: water activity a_w as a function of the mole fraction of water X_w for solutions of glucose and fructose. Adapted from Audu et al. (1978).

diffusion of reactants may become increasingly difficult when the water content decreases. Another reason is that activity coefficients of reactants are affected by the presence of cosolutes, the concentrations of which increase if water is removed. A rather fundamental and useful overview of water activity in relation to chemical reactions has been given by Blandamer *et al.* (2005). Nevertheless, the food stability map is useful as rule of thumb.

For a decade or so, interest has shifted to the effect of glass transition in which water also plays a prominent role. How are glasses formed? Suppose that a solution is cooled and that at the crystallization temperature T_m the solutes remain in solution because cooling is done quickly and there is not enough time for crystal nuclei to be formed. Glasses are formed when the supersaturated solution solidifies eventually at the so-called glass transition temperature T_g. This transition can be detected via a change in heat capacity and measured via DSC (differential scanning calorimetry). In the case of foods this is usually a glass transition temperature range, rather than one specific temperature. Such a transition is characterized by an enormous increase in viscosity when an amorphous matrix is formed. As a rule of thumb, the viscosity η_g at T_g is around 10^{12} Pa s. The glass transition is the manifestation of these drastic changes in molecular mobility. We will not discuss all the intricacies of the glassy state and how it is characterized; references include Lievonen *et al.* (1998), Roos (1995) and Roos *et al.* (1996), Le Meste *et al.* (2002). Foods in the glassy state usually have a high stability and a long shelf life because of the fact that the molecular mobility is so low. The molecular mobility strongly depends on T-T_g, in other words how much the actual temperature deviates from the glass transition temperature. It should be recognised however that glassy foods are in a non-equilibrium state, and therefore there is an inherent tendency to change, albeit at an infinitely slow rate, at least in principle. In practice however, the molecular mobility of water especially is not absolutely zero, and neither is that of small solutes. The glass transition temperature is strongly dependent on composition and especially the water content. Water can act as a plasticizer, or in other words, the viscosity may increase drastically at a certain water content, changing from the amorphous glassy state into a supercooled, viscous or rubbery state. Or stated another way, when the water content increases, the glass transition temperature decreases. When water acts as a plasticizer, it leads to drastic changes in the mechanical properties and stability of the food, sometimes referred to as collapse of the matrix, and causing stickiness. Figure 4.3 gives a very schematic impression of the changes in rates of quality loss as a function of temperature in the case of a glass transition range. The effective diffusion coefficient is almost zero for the compounds forming the glass. However, small molecules such as water and oxygen are still able to diffuse, albeit slowly.

As for microbial stability, it goes without saying that micro-organisms are omnipresent and are a constant threat to the food and its consumer. In fact, microbial food safety is the first and foremost thing to consider in food design. The growth

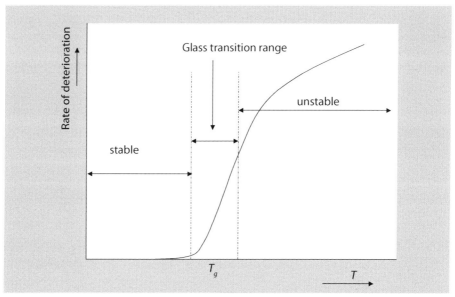

Figure 4.3. Highly schematic representation of the rate of quality loss in a food undergoing a glass transition.

of micro-organisms depends notably on food composition (pH, water activity) and temperature. Hygiene is very important (see the chapter on hygienic design in this book): if the raw materials already contain large amounts of micro-organisms there is going to be a problem with the resulting product. Processing can help to reduce the number of micro-organisms, but the current trend of minimal processing gives reason for concern, and only emphasizes the need for hygiene in raw materials, and critical analysis at each part of the food chain. Micro-organisms are of the utmost importance in foods for three reasons: i) they can be pathogenic, ii) they can spoil foods, iii) they can be used in food production via fermentation. Food microbiology is a topic in its own right, and this is not the place to discuss it in great detail. In section 4.3 we will briefly discuss growth models and inactivation kinetics for micro-organisms.

The above short description of key reactions in foods demonstrates that it can be quite complicated to manage all this, all the more so because of interactions between reactions. For instance, a change in pH due to one reaction can have an effect on the course of another reaction. One way to get a better grip on quality changes and product design is to use models.

4.3. Types of models

Models are abstractions of the real world, and are only approximations of reality: 'all models are wrong, but some are useful' is a famous quotation from the statistician George Box (1976). One should be aware of their limitations, but also of their usefulness. For instance, a growth model for *Salmonella* in poultry may not be directly applied to growth of the same *Salmonella* in pork meat. Often it is dangerous to extrapolate a model outside the region for which it was validated. However, the big advantage of models is that one can simulate and predict certain properties, and it goes without saying that this is very useful for product and process design. The use of models can save time and money, improve a design considerably, increase productivity, and shorten the time to market (Levine, 1997). Modelling may greatly reduce experimentation, but it should be realised that a certain amount of experimentation is always necessary, in order to derive relevant parameters and validate models. However, when done in the right way, modelling will make experimentation much more efficient because one then knows how experiments should be set up.

A model is thus in very general terms a simplified description of parts of the real world. A conceptual model, for instance, depicts relationships in a systematic but qualitative way; it is a hypothesis about how a system works and how it responds to changes in inputs. A mathematical model on the other hand depicts relationships in a quantitative way via one or more equations. Mathematical models can be subdivided into empirical and mechanistic models. Empirical models give a mathematical description of some relationship without any theoretical background; mechanistic models are built upon a theoretical concept, for instance, a known chemical or physical mechanism. Mechanistic models are preferable, because they are more reliable for making predictions and extrapolations. Empirical models may never be extrapolated outside the region for which they were derived. Since foods are so enormously complex and many phenomena are not fully understood, we often have to resort to empirical models in food science. An important example is predictive microbiology where the models used are empirical, or semi-empirical at best, because there is not yet a complete mechanistic description of the growth of micro-organisms in foods.

Another way of classifying models is to differentiate between deterministic and stochastic models. Deterministic models always give the same output with the same input. However, the real world is not deterministic; there is always variability and uncertainty. This uncertainty can also be modelled, and a model that does this is called a stochastic model. This brings us to the topic of statistics. In order to model uncertainty and variability the use of statistics is indispensable. Statistics helps in interpreting data by separating noise from relevant information; it helps in setting up experiments and it introduces the stochastic part to express uncertainty in models. In short, statistics is the science of learning from experience (Box, 1976).

4.4. Modelling of food quality attributes

Mathematical models are built to control and predict certain properties of foods. They consist of equations that provide an output (e.g. vitamin content) based on a set of input data (e.g. time, temperature). It is a concise way to express physical behaviour. Research in the area of predictive modelling has been booming, especially in the field of microbiological growth models. Such models are of great help in predicting the safety and shelf life of foods in relation to microbiological problems. Modelling as such is not new in food technology. One of the earliest models was developed in the 1920's to predict the inactivation of micro-organisms as a function of heating time and temperature, the so-called Bigelow model. This model has been of great help in optimising processes for the sterilization of foods, especially in the canning industry. Every food technologist is familiar with the D and Z values that are used in this model. Incidentally, this model has been criticized recently and a new model has been proposed (Van Boekel, 2002, 2008).

Modelling food quality attributes means modelling changes: the quality of a food nearly always changes over time. Food quality modelling is therefore almost synonymous with kinetic modelling (Van Boekel, 2008). The consequence is that differential equations frequently form the basis for mathematical models; these can sometimes be solved analytically, but if not it is relatively easy nowadays to solve them numerically with the available software, or even using spreadsheets. When it comes to food quality attributes, a subdivision can be made into chemical, biochemical, physical and microbial phenomena, as indicated above under key reactions. A few examples of each will be given below.

4.4.1. Modelling chemical reactions

Suppose we have a reaction between two molecules A and B, which yield two products P and Q:

$$A + B \rightarrow P + Q \tag{4.2}$$

The rate is then defined as:

$$-\frac{d[A]}{dt} = -\frac{d[B]}{dt} = \frac{d[P]}{dt} = \frac{d[Q]}{dt} = k[A][B] \tag{4.3}$$

The proportionality constant k is the so-called rate constant. For molecules to react, they must first come together, and this happens via diffusion. If the encounter frequency is rate limiting, a reaction is called diffusion-limited, which implies that the reaction itself takes place very rapidly. This is the case for acid-base reactions and radical reactions, for instance. The rate constant for such a case is:

$$k_{dif} = \frac{8 \cdot 10^3 \, RT}{3\eta} \quad (dm^3.mol^{-1}.s^{-1}) \tag{4.4}$$

R is the gas constant (J mol^{-1} K^{-1}), T absolute temperature (K), η the viscosity of the solution (Pa s) . If we take Equation (4.4) as the measure for the fastest bimolecular reaction possible, it is found that for η = 1 mPa s (viscosity of water at 20 °C) k_{dif}=6.6×10^9 dm^3 mol^{-1} s^{-1} and at 100 °C k_{dif}=3×10^{10} dm^3 mol^{-1} s^{-1}. These should be roughly the upper limits for bimolecular reaction rate constants in aqueous solutions at the temperature indicated. The effect of temperature on the encounter rate is incorporated via the effect of temperature on the viscosity of the solvent.

In most cases, however, the actual reaction step will be rate-limiting rather than the encounter rate. Instead of using Equation (4.3) we move to the most simple equation possible, the so-called general rate law, which is for a single reactant at concentration c:

$$r = -\frac{dc}{dt} = kc^n \tag{4.5}$$

This differential equation is thus in the form of a power law expression, where n is the so-called order of the reaction. The equation reflects the dependence of rate r on concentration for just one component; k is again the reaction rate constant. The unit for k for a reaction having order n is: (dm^3 mol^{-1})$^{n-1}$ s^{-1}. Equation (4.5) can be integrated with respect to time to obtain the course of the concentration as a function of time:

$$c^{1-n} = c_0^{1-n} + (n-1)kt \quad \text{for } n \neq 1 \tag{4.6a}$$

$$c = c_0 \exp(-kt) \text{ for } n = 1 \qquad\qquad (4.6b)$$

c_0 is the initial concentration at $t = 0$.

Figure 4.4 shows a decomposition reaction for dimensionless scales and varying order, using Equation (4.6). It appears that no real distinction can be made between the models if the fractional conversion is less than, say, 20-30%. In other words, for a proper estimation of the order, one should conduct the experiment such that a considerable extent of reaction is reached. Proper experimental design is therefore of the utmost important; in this case it is the product $k{\cdot}t{\cdot}c_0^{n-1}$ that determines the extent of the reaction. If $t > 1/((1-n)c_0^{n-1}k)$ for $n < 1$, $c_t/c_0 = 0$ whereas for $n \geq 1$ c_t/c_0 approaches 0 asymptotically. It should also be noted that in a closed system a reaction order $n=0$ cannot run indefinitely; the order will have to change at some point in time.

In food science literature, quality changes are usually modelled via a zero-, first-, or second-order reaction. From Equation (4.6a) it follows that for $n = 0$ for a decomposition reaction:

$$-\frac{dc}{dt} = k \qquad\qquad (4.7a)$$

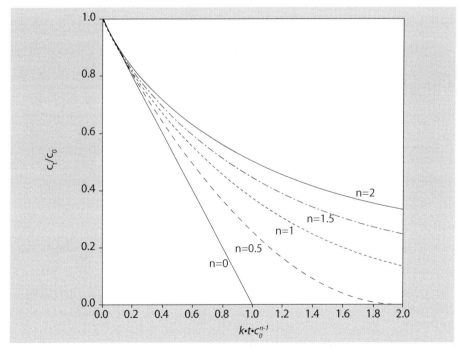

Figure 4.4. Decomposition of a component for a reaction having the same initial concentration and rate constant but a varying order n.

Integration leads to:

$$c = c_0 - kt \qquad (4.7b)$$

Zero-order reactions are rather frequently reported for changes in foods, especially for formation reactions when the amount of product formed is only a small fraction of the amount of precursors present, or for decomposition reactions where only a small amount of product is formed from a reactant. The reactant is then in such large excess that its concentration remains effectively constant throughout the observation period, and hence the rate appears to be independent of the concentration. A frequently reported example of a zero-order reaction is the formation of brown colour in foods as a result of the Maillard reaction: see Figure 4.5.

The kinetics of Maillard-type browning is rather intricate, and it is just fortuitous that a zero-order reaction equation fits. A much more detailed analysis of the kinetics of the Maillard reaction is given by Martins and Van Boekel (2004, 2005b).

First-order reactions are also frequently reported for reactions in foods. The equations for a degradation reaction for $n = 1$ are:

$$-\frac{dc}{dt} = kc \qquad (4.8a)$$

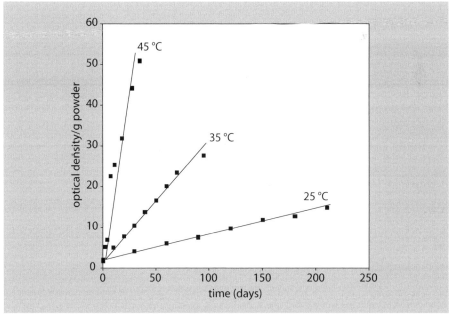

Figure 4.5. Example of a zero-order reaction reported for the non-enzymatic browning of whey powder. Adapted from Labuza (1983).

Integration leads to:

$$c = c_0 \exp(-kt) \tag{4.8b}$$

Frequently, the logarithmic form is used instead of the exponential equation:

$$\ln c = \ln c_0 - kt \tag{4.8c}$$

The nonlinear Equation (4.8b) is thus transformed into the linear Equation (4.8c). An example of a food-related first-order reaction is shown in Figure 4.6. It concerns the heat-induced degradation of betanin, a natural colour compound from red beets. Figure 4.6a shows the first-order plot for untransformed data according to Equation (4.5b), while Figure 4.6b shows the plot for logarithmically transformed data according to Equation (4.5c). A log plot resulting in a straight line is frequently taken as proof of a first-order reaction. The plot in Figure 4.6b indeed looks reasonably straight. While this may be done for a visual check, such a transformation should not be performed for estimating the rate constant, for statistical reasons. The problem is that upon transformation not only the data are transformed but also the error structure related to the data, and this may lead to bias in estimation (Van Boekel, 1996, 2008).

To show the problems of transformation, unfortunately widely applied, we give an example with real data on the degradation of carotenoids in olives, which was claimed to be a first-order degradation. Figure 4.7a shows a linearized first-order plot for violaxanthin with the linear least squares regression results. The fit is actually quite

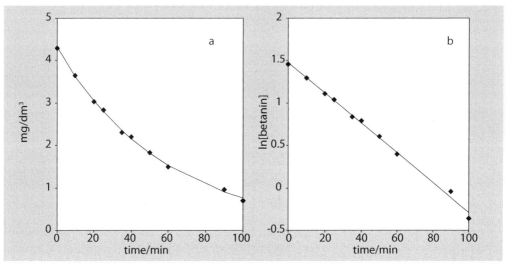

Figure 4.6. Example of a first-order reaction for the degradation of betanin at 75 °C. (a) Untransformed concentration and the fit according to a first-order reaction, Equation (4.5b) and (b) log-transformed data fitted by a linear line according to Equation (4.5c). Adapted from Saguy et al. (1978).

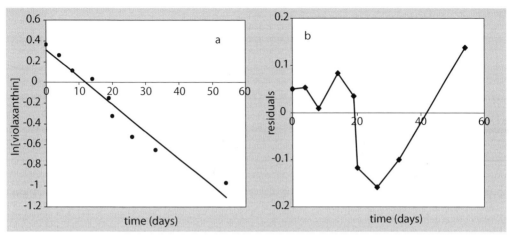

Figure 4.7. (a) First-order linearized plot for degradation of violaxanthin in olives in mg/kg (logarithmic data) as a function of time. The line is obtained by linear regression: y=0.314 – 0.0263x (r^2=0.937). (b) Residuals. Adapted from Mínguez-Mosquera and Gandul-Rojas (1994).

bad when judged by the residuals (Figure 4.7b), which show a strong trend; residuals are obtained from the difference between measured and modelled values. Residuals should be spread randomly and not show a trend. Incidentally, the coefficient of determination $r^2 = 0.937$ seems to indicate quite a good fit. However, r^2 is not a good parameter to indicate goodness of fit, nor linearity. The correlation coefficient r measures how 2 sets are (linearly) related, but it does not prove linearity or adequacy of fit. It is in this respect a widely misused parameter, and there is not much sense in reporting its value to indicate goodness of fit. Residuals are a much better check. In any case, a significant correlation should not be taken as indication for causality, so don't be fooled by a high correlation coefficient.

One of the possibilities for the bad fit in Figure 4.7 is the logarithmic transformation (another possibility is that a first-order model does not apply). Let us see what happens if we do not transform the data, and perform nonlinear regression: see Figure 4.8.

The nonlinear regression fit appears to be better in terms of residuals, although it is certainly not perfect. It may therefore be that a first-order model is not the best model. In any case, the estimated rate constant obtained via linear regression is biased (as well as the estimation of the initial concentration).

Second-order kinetics is reported in food science literature not as frequently as one might perhaps expect. A likely reason is the following. If we take Equation (4.2) and suppose that one of the reactants, say B, is present in excess, it follows that:

$$-\frac{d[A]}{dt} = k[A][B] = k'[A] \qquad (4.9a)$$

$$k' = k[B] \qquad\qquad\qquad (4.9b)$$

k' is called a pseudo first-order rate constant, which is constant as long as [B] does not change notably. This goes to show that the experimentally observed kinetics of a reaction does not necessarily correspond to the actual mechanism. The equation for a second-order reaction, $n = 2$, is:

$$-\frac{dc}{dt} = kc^2 \qquad\qquad\qquad (4.10a)$$

Integration leads to:

$$c = \frac{c_0}{1 + c_0 kt} \qquad\qquad\qquad (4.10b)$$

In its linearized form it reads:

$$\frac{1}{c} = \frac{1}{c_0} + kt \qquad\qquad\qquad (4.10c)$$

Second-order reactions are sometimes reported for changes to amino acids involved in the Maillard reaction. A case in point is the loss of lysine (bound in proteins, hence the ε-amino group of lysine) in sterilized milk due to the Maillard reaction. According to literature this is a second-order reaction in lysine, i.e. a plot of the inverse of [lysine] versus time gives a straight line (Equation 4.10c): see Figure 4.9. The actual mechanism of lysine loss is much more complicated than a relatively simple bimolecular reaction: apart from the initial condensation with lactose, there is regeneration of lysine (it acts as a catalyst) but subsequent further reaction of lysine

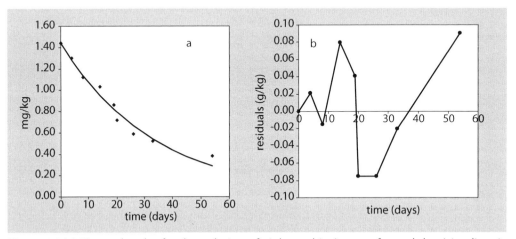

Figure 4.8.(a) First-order plot for degradation of violaxanthin (untransformed data) in olives in mg/kg as a function of time. The solid line is obtained via nonlinear regression of the first-order model: y=1.44exp(-0.0297x) (b) Residuals. Same data as in Figure 4.7.

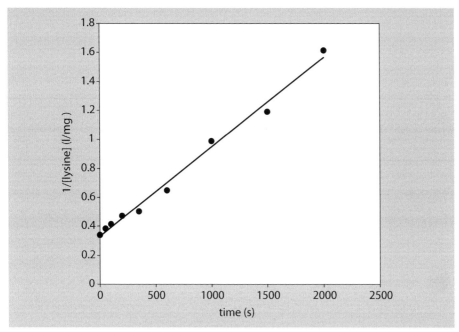

Figure 4.9. Lysine loss in milk heated at 160 ° C plotted as 1/[lysine] versus time according to a second order model (drawn line). Adapted from Horak (1980).

residues also occurs with intermediate and advanced Maillard reaction products (Brands and Van Boekel, 2002).

Even though simple kinetics are frequently applied in Food Science, reactions are usually much more complicated. Examples are fat oxidation and the Maillard reaction; they cannot be given in one simple equation; rather so-called multiresponse modelling is needed (Brands and Van Boekel, 2002; Martins and Van Boekel, 2005a,b; Van Boekel, 1999, 2000, 2008). This is not discussed any further here; the reader is directed to the references given.

In many cases, it may actually happen that more models apply to a given reaction. The question is then how to discriminate between various models. A statistical analysis is very helpful in this case. We give one example on the colour change in olives during fermentation. The colour of olives is to a large extent determined by carotenoids, and during fermentation these carotenoids are not stable as a result of which the colour changes. These changes were studied in detail by Mínguez-Mosquera and Gandul-Rojas (1994). The question now is which model can be used based on these data. A first and second-order model were applied to model the degradation of the carotenoid violaxanthin. Figure 4.10 shows both models for the dataset of violaxanthin. A statistical analysis is now needed to decide which model performs best. An ANOVA table is quite helpful in this respect, as reproduced in Table 4.3.

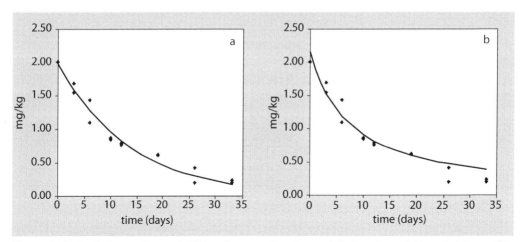

Figure 4.10. Fit of a first-order model (A) and a second-order model (B) to degradation of violaxanthin in olives. Adapted from Mínguez-Mosquera and Gandul-Rojas (1994).

Table 4.3. ANOVA table for regression analysis of a first- and second-order model to degradation of violaxanthin in olives.

	First- order model				Second-order model			
	Sum of squares	Degrees of freedom	Mean square	F-value	Sum of squares	Degrees of freedom	Mean square	F-value
Regression	4.301	1	4.301	369.4	4.192	1	4.192	209.2
Error	0.151	13	0.0116		0.261	13	0.020	
Lack of fit	0.058	6	0.0097	0.73	0.167	6	0.028	2.09
Pure error	0.093	7	0.0133		0.093	7	0.013	
Total	4.4524	14			4.452	14		

The critical F-value for the indicated degrees of freedom is 3.87 at the 95% confidence level. Although the lack-of-fit F-value for the second-order model is considerably higher than that for the first-order model, it is still below the critical value, and hence also the second-order model passes the test for goodness of fit. Unless a lack of fit is immediately apparent, more sophisticated tests for model discrimination are needed. One such test is the so-called Akaike criterion (Burnham and Anderson, 1998):

$$AIC_c = n\ln\left(\frac{SS_r}{n}\right) + 2(p+1)\left(\frac{n}{n-p}\right) \tag{4.11}$$

in which n is the number of data points, SS_r the residual sums of squares, and p the number of parameters. The model with the lowest number performs the best (in a statistical sense). In this case, the AIC_c for the first-order model is -21.8 and for the second-order model it is -18.2, hence the first-order model is preferable.

Table 4.4. Regression results for the parameters describing the degradation of violaxanthin in olives using the first-order model shown in Figure 4.10a.

Parameter	Point estimate	95% confidence interval
k	0.072	0.011
c_0	1.98	0.16

Another area in which statistics are needed is in evaluating the precision of the parameter estimates. In Equations (4.3) and (4.6), for instance, the rate constant k and the initial concentration c_0 are the parameters, to be estimated via regression (in this case via nonlinear regression). Along with a numerical value of the parameters it is also essential to know the precision with which these values are obtained. Table 4.4 gives the result for this example for the first-order model; it makes no sense to do this for the second-order model because this was considered less suitable as shown above. The precision of parameters can be given in two ways, namely as standard deviation and as 95% confidence intervals. The latter is preferable because it includes the degrees of freedom and hence gives a better impression of the real precision obtained (the more degrees of freedom the better). A confidence interval (CI) and the standard deviation (SD) are linked in the following way:

$$CI = \pm SD \cdot t_{(1-0.5\alpha),v} \tag{4.12}$$

$t_{(1-0.5\alpha),v}$ is the t-statistic for a confidence level ($1-0.5\alpha$, for a 95% CI, $\alpha=0.05$) and degrees of freedom v $(=n-p)$. Ultimately, numerical values of parameters are needed to perform subsequent calculations so that predictions can be made. It may be that estimates are rather imprecise, and in that case it does not make much sense to use such estimates for predictions because they will then be very imprecise and hence useless. If this happens, more effort must be made to obtain more precise estimates, either by taking more measurements (increasing n), or more precise measurements. For a more detailed account of the importance of using statistics in modelling the reader is referred to Van Boekel (1996, 2008).

4.4.2. Modelling temperature dependence of chemical reactions

Arrhenius' law was empirically derived to describe the temperature dependence of chemical reactions. It has proven to be very worthwhile in chemical kinetics. It relates the rate constant k of a reaction to absolute temperature T:

$$k = A\exp(-\frac{E_a}{RT}) \tag{4.13a}$$

The linearized form is:

$$\ln k = \ln A - \frac{E_a}{RT} \tag{4.13b}$$

in which A is a so-called 'pre-exponential factor' (sometimes called the frequency factor), and E_a the activation energy, and R and T the gas constant and absolute temperature, respectively. The dimension of A should be the same as that of the rate constant k; it therefore does have units of frequency only in the case of a first-order reaction. The activation energy can be seen as the energy barrier that molecules need to cross in order to be able to react. The proportion of molecules able to do that increases with temperature, which qualitatively explains the effect of temperature on rates. Arrhenius' equation gives a quantitative account. The physical meaning of A is that it represents the rate constant at which all molecules have sufficient energy to react (i.e. $E_a = 0$). Incidentally, it is not a good idea to derive the activation energy parameters from linear regression of $\ln k$ vs. $1/T$ because of the transformation of data points with their errors by taking logarithms; rather, non-linear regression should be used, as discussed above. Another remark in this respect is that the two-step procedure of first deriving rate constants and then regressing them versus temperature usually results in very wide confidence intervals if only 3-4 temperatures have been studied, as is frequently the case. A better approach is to substitute the rate constant in the appropriate rate equations and perform a non-linear regression (Van Boekel, 1996, 2008). For instance, for a first-order reaction this would be:

$$c = c_0 \exp\left(-A\exp(-\frac{E_a}{RT})\cdot t\right) \tag{4.14}$$

In this way, all data are used at once to estimate the activation parameters and a much more precise estimate of these parameters is obtained. It probably remains a good idea to present Arrhenius' expression in the form of a plot of $\ln k$ or $\ln(k/T)$ vs. $1/T$ because any deviation of the data from these expressions becomes immediately apparent. In doing so, however, the values of the parameters estimated by non-linear regression should be used to construct the plot. The first step should always be to check the validity of the law of Arrhenius, and only if it appears to be correct should the next step be the estimation of the activation parameters. Obvious as this may seem, this rule is not always obeyed. It is essential to realize that the concept of activation energies is strictly speaking only valid for elementary reactions.

It is possible to reparameterize the Arrhenius' equation; and it is actually desired from a statistical point of view (Van Boekel, 1996, 2008). A very simple reparameterization is to introduce a reference temperature T_{ref}. The basis for this arises from the application of Equation (4.13a) at two temperatures T_1 and T_2:

$$k_1 = A\exp(-\frac{E_a}{RT_1}) \qquad \text{(based on 4.13a)}$$

$$k_2 = A\exp(-\frac{E_a}{RT_2}) \qquad \text{(based on 4.13a)}$$

If one arbitrarily chooses a reference temperature, say $T_2=T_{ref}$, one can combine these two equations, assuming that the pre-exponential factor and E_a do not depend on temperature:

$$\frac{k}{k_{ref}} = \exp\left(-\frac{E_a}{R}\left(\frac{1}{T} - \frac{1}{T_{ref}}\right)\right)$$ (4.15)

The actual result of this is that the pre-exponential factor is replaced by a rate constant at some reference temperature k_{ref}. The reference temperature should preferably be chosen in the middle of the studied temperature regime.

There are also other Arrhenius-like equations proposed in literature that could be used just as well, but which are not commonly used. The following equation would do equally well as the Arrhenius' equation:

$$k = A\exp\left(-\frac{B}{T}\right)$$ (4.16)

with A and B as fit parameters without a physical meaning. Although this seems undesirable, one has to realise that sometimes parameters do not really have a physical meaning. For instance, if one determines an activation energy for microbial inactivation, what does it mean if an activation energy of, say 300 kJ/mol, has been derived? A mole of bacteria is somewhat hard to envisage. It would actually be better to use Equation (4.16) for phenomena that do show Arrhenius-like behaviour but do not really reflect a defined chemical reaction. For instance, the effect of temperature on diffusivity can often be described using the Arrhenius' equation, but there is no activation energy for molecular mobility (though there may be barriers), and therefore it does not make much sense to report an activation energy for diffusion; temperature coefficients A and B like in Equation (4.16) seem more appropriate.

It is perhaps instructive to consider the range of values that rate constants can take. Table 4.5 shows orders of magnitude for rate constants, depending on conditions. The very large effect of activation energy on the rate of a reaction is apparent. In fact, without activation barriers, reactions would be so fast that foods would spoil immediately.

Table 4.5. Orders of magnitude for rate constants of bimolecular reactions in aqueous solutions at 25 °C.

Conditions	Order of magnitude of k (dm^3 mol^{-1} s^{-1})
No diffusion limit and no barrier[1]	10^{14}
Diffusion limit, no activation energy[2]	10^{10}
No diffusion limit: activation energy 25 kJ/mol	10^{10}
No diffusion limit: activation energy 50 kJ/mol	10^5
No diffusion limit: activation energy 100 kJ/mol	10^{-4}

[1] This is in fact the value of the pre-exponential factor in the Arrhenius' equation, corresponding to the hypothetical situation that $T \rightarrow \infty$.
[2] As given by Equation 4.4.

The parameters that have been discussed so far, orders, rate constants, activation parameters, etc., are actually all that is needed in (chemical) kinetics. Unfortunately, it has become the habit to use several other kinetic parameters in food science. They originate from days gone by when it was necessary to derive parameters and models to describe (mainly microbial) changes in foods during processing and storage when it was not possible to make use of modern reaction kinetics. Useful as they have been, we would like to discourage further use of these parameters in food science. They do not provide extra information and may even cause confusion; in some instances they are less reliable than the underlying fundamental parameters. Fortunately, all these parameters are related to the more fundamental parameters that we have discussed so far. Although we discourage further use, we give a brief overview of these parameters so that the reader can see how they relate to the fundamental parameters discussed above.

The parameter Q_{10} describes the temperature dependence of a reaction as the factor by which the reaction rate is changed when the temperature is increased by 10 °C:

$$Q_{10} = \frac{k_{T+5}}{k_{T-5}} \approx \frac{k_{T+10}}{k_T} \qquad (4.17)$$

If the Arrhenius' equation holds, it can be shown that:

$$Q_{10} = \exp\left(\frac{10E_a}{RT^2}\right) \qquad (4.18)$$

The parameter is thus seen to depend strongly on temperature, which is a drawback.

Another parameter to describe temperature dependence is Z, which expresses the increase in temperature that would produce an increase in rate of a factor 10. Z is defined as:

$$Z = \frac{2.303RT^2}{E_a} = \frac{10}{\log Q_{10}} \qquad (4.19)$$

Like the parameter Q_{10}, Z is temperature dependent, which restricts its use. Z is frequently used in bacteriology to describe inactivation of micro-organisms.

Also used is the parameter D, especially in thermobacteriology. It is the decimal reduction value, the time needed to reduce a concentration by a factor 10. D is nothing other than an inverse rate constant. For a first-order reaction:

$$D = \frac{\ln 10}{k} = \frac{2.303}{k} \qquad (4.20)$$

and for a second-order reaction:

$$D = \frac{9}{c_0 k} \qquad (4.21)$$

A plot of D versus T' (in °C) is usually taken to be a straight line (for a limited temperature range), see Figure 4.11. D relates to the Z value, as k is related to E_a:

$$\log D = \log D_R - \frac{T'-T'_R}{Z} \tag{4.22}$$

D_R is the reference value of D at the reference temperature T'_R (often chosen as 250 °F for historical reasons, which is equal to 121.1 °C). Equation (4.22) is referred to as the TDT curve (thermal death time curve) or the Bigelow model.

As shown, all these parameters can be represented by the more fundamental kinetic parameters and, in our opinion, there is no real advantage in using them anymore.

4.4.3. Modelling biochemical reactions

Biochemical reactions are important for food quality, as mentioned in section 4.2, because most foods, being biological materials, contain enzymes. Sometimes these enzymes are desired (for instance, in cheese ripening) but mostly enzymes need to be deactivated because otherwise their action will lead to deterioration of food quality. Examples are the enzymatic browning of apples, potatoes, and cauliflower due to polyphenoloxidase, or formation of a soapy taste in raw milk due to the action of lipase. If one wants to exploit enzymes, enzyme kinetics is useful. In most cases reported in literature, Michaelis-Menten kinetics is applied, although one

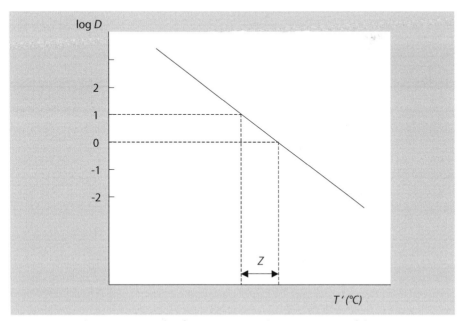

Figure 4.11. Schematic example of a TDT curve and interpretation of the Z value.

should check whether or not Michaelis-Menten kinetics is actually applicable. The Michaelis-Menten equation reads:

$$v = v_{max} \frac{[S]}{[S] + K_M} \qquad (4.23)$$

in which v is the *initial* rate of the reaction, v_{max} the maximum rate of the enzyme under the conditions studied, [S] is the substrate concentration, and K_M the Michaelis constant. v_{max} and K_M are the parameters of the equation. With knowledge of these parameters the rate of the enzymatic reaction can be predicted. Frequently, Lineweaver-Burke plots are made to estimate the kinetic parameters, but this should not be done because of transformation of errors, as discussed above. Nonlinear regression estimation is preferable. Figure 4.12 gives a simple example of Michaelis-Menten kinetics.

An extensive overview of enzyme kinetics can be found in Marangoni (2003) and Van Boekel (2008).

If one wants to prevent the action of enzymes, inactivation kinetics is needed. Enzymes are proteins and inactivation is due to unfolding of the protein. A general model for that is:

$$N \underset{k_2}{\overset{k_1}{\rightleftharpoons}} D \overset{k_3}{\longrightarrow} I \qquad (4.24)$$

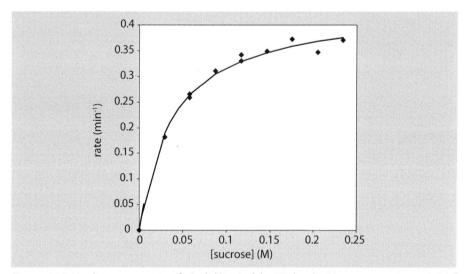

Figure 4.12. Nonlinear regression fit (solid line) of the Michaelis-Menten equation to initial rates of sucrose hydrolysis by yeast invertase as a function of substrate concentration. Adapted from Chase et al. (1962).

in which N represents the native protein, D the denatured protein and I the inactivated protein, and the *k* values represent the rate constants for each step. The importance of the equilibrium between N and D is that proteins can refold after denaturation, and hence enzyme activity may be restored upon removing the cause of denaturation. In most cases in foods, the cause of denaturation is heating. In any case, if the denatured protein is subject to further reactions leading to the inactive form I, the enzyme cannot return to its active form, and consequently enzyme activity is lost. In most cases, a first-order model as given in Equation (4.6b) appears to be applicable to describe enzyme inactivation. This implies that the third step in Equation (4.24) is rate-determining. Many examples of inactivation curves can be found in food science literature. Figure 4.13 shows an example in which first-order behaviour is apparent. Figure 4.14 on the other hand shows an example of biphasic behaviour, which can also be frequently observed, so one should not automatically assume that enzyme inactivation is by definition first-order behaviour.

4.4.4. Modelling physical reactions

Physical reactions frequently lead to quality change. Examples are creaming or sedimentation, fracture phenomena, viscosity changes, gelation of biopolymers, crystallisation, and moisture migration. Modelling these phenomena is not easy because the changes are rather complex, and may be accompanied by chemical changes. As an example we present two models for predicting viscosity of dispersions. The first one is an equation derived by Einstein for dilute dispersions:

$$\frac{\eta}{\eta_s} = 1 + 2.5\varphi \qquad (4.25)$$

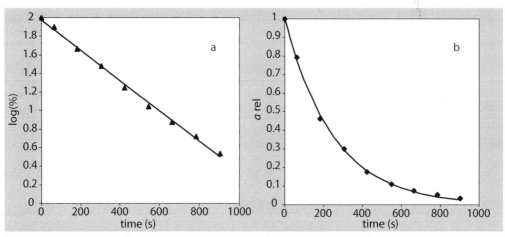

Figure 4.13. An example of apparent first-order heat-induced inactivation kinetics of pectin-methylesterase from tomato at 69.8 °C, presented as a logarithmic plot (a) and as relative activity plot (b). The lines represent a first-order model. Adapted from Anthon et al. (2002).

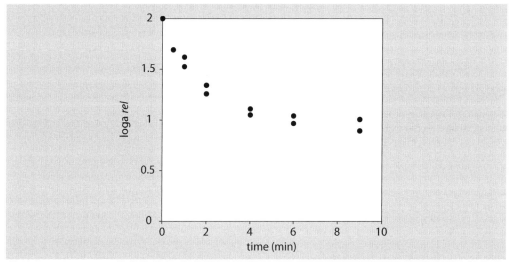

Figure 4.14. Logarithmic plot showing biphasic inactivation of the enzyme cathepsin D in milk at 62.6 °C. Adapted from Hayes et al. *(2001).*

in which η represents the viscosity of the dispersion, η_s the viscosity of the solvent, and φ the volume fraction of the dispersed particles. The interesting aspect of this equation is that only the volume fraction but not the size of the dispersed particles is of importance in determining the viscosity. However, this equation is only valid for very dilute dispersions ($\varphi < 0.01$), and therefore not very suitable for foods. For more concentrated dispersions, an empirical relation has been derived, which is the so-called Eilers equation:

$$\frac{\eta}{\eta_s} = \left(1 + \left(\frac{1.25\varphi}{1 - \dfrac{\varphi}{\varphi_{max}}} \right) \right)^2 \tag{4.26}$$

This equation works quite well for foods. Anema *et al.* (2004) applied this model to skimmed milk samples with varying volume fraction of casein micelles. Manski *et al.* (2007) used such a model to describe the influence of dispersed particles on deformation properties of concentrated caseinate composites. Such equations can thus be used to predict the rheological properties of a food if one knows the volume fraction of dispersed particles. As indicated above, there are numerous models describing physical phenomena, such as aggregation and flocculation, crystallization kinetics, drying and dehydration. A useful reference for all kinds of physical models is to be found in Walstra (2003).

4.4.5. Modelling microbial reactions

Microbiological changes are due to the growth of micro-organisms. This is sometimes desirable in fermentation, but usually undesirable in other environments because microbial growth may lead to spoilage and even health-threatening situations when pathogens come into play. Regardless of this fact, the ability to predict growth of bacteria in foods is of the utmost importance for food design and predicting shelf life. A frequently used growth model is the modified Gompertz model:

$$\ln \frac{N}{N_0} = A_s \exp\left[-\exp\left(\frac{\mu_{max} e}{A_s}(\lambda - t) + 1\right)\right] \tag{4.27}$$

in which N is the number of micro-organisms, N_0 the number of micro-organisms at time zero, A_s is the asymptotic value of the maximum number of micro-organisms, μ_{max} the maximum growth rate, λ the lag phase, and e is the number 2.718 (=exp(1)). Figure 4.15 shows this graphically.

Figure 4.16 shows an example of the modified Gompertz model applied to the growth of Salmonellae in a laboratory medium. The parameters are in this case: A_s=13.14, λ=4.37 h, μ_{max}=0.70 h^{-1}.

It should be noted that there are many more growth models published. A quick scan of the *International Journal of Food Microbiology, Journal of Food Protection,* and *Food Microbiology* will overwhelm the reader with the many variations on a theme. However, a useful overview of the state of the art is McKellar and Lu (2004) and Brul *et al.* (2007).

Another important aspect related to microbiology is the ability to inactivate micro-organisms in foods, and for that we need inactivation kinetics. As mentioned above, the first model published in food science was on this topic in the 1920's, and was the

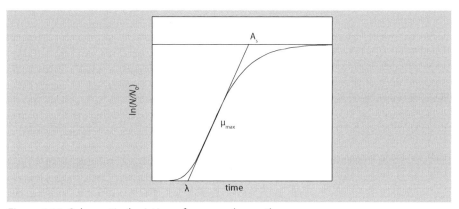

Figure 4.15. Schematic depiction of a general growth curve.

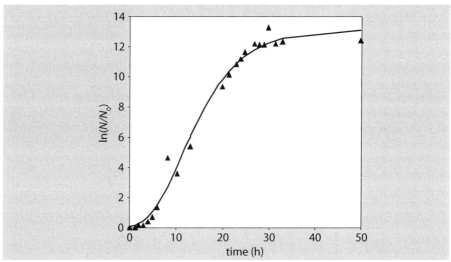

Figure 4.16. Example of the fit of the modified Gompertz equation to the growth of Salmonellae in a laboratory medium. Fit parameters: A_s = 13.14, λ = 4.37 h, μ_{max} = 0.70 h^{-1}. Adapted from Gibson et al. (1988).

so-called Bigelow model still used today. It is actually a first-order model as displayed in Equation (4.8b):

$$S(t) = \exp(-\frac{t}{D}) \tag{4.28}$$

or

$$\log S(t) = -\frac{t}{D} \tag{4.29}$$

in which

$$S(t) = \frac{N}{N_0} \tag{4.30}$$

with D the decimal reduction time already displayed in Equation (4.20). This model is widely applied in food science, probably because it is so simple. Nevertheless, there are problems with it. Equation (4.29) shows that a plot of $\log S(t)$ versus time should be linear, but if one screens the literature one will find that most plots are surprisingly nonlinear. This fact is simply ignored by many authors. In order to account for this nonlinearity, a new model has come up (Peleg and Cole, 1998), which has been tested for a number of cases by Van Boekel (2002). The model is a so-called Weibull model:

$$S(t) = \exp\left(-(\frac{t}{\alpha})^{\beta}\right) \tag{4.31}$$

and

$$\log S(t) = -\left(\frac{t}{2.303\alpha}\right)^{\beta} \qquad\qquad (4.32)$$

In comparison with Equations (4.28) and (4.29) one extra parameter is added, namely β. This is the so-called shape factor because its value determines the shape of the inactivation curve (two examples are given below). The parameter α has units of time and could be considered as the alternative for the D value. The interesting aspect is that if β = 1, then the Weibull model reduces to the Bigelow model; it is thus a rather flexible model. However, there are only a few cases in which β=1 (Van Boekel, 2002). Figures 4.17 and 4.18 each show an example of the fit of the Weibull model to the inactivation of micro-organisms.

4.5. Applications of models to reactions in foods

4.5.1. Exploiting differences in temperature dependencies

When the effect of temperature on reactions in foods has been established, preferably in the form of the parameters discussed, i.e. activation energy and pre-exponential factor, the value of the parameters needs some discussion. Occasionally, there seems to be some misunderstanding regarding interpretation. For instance, if a high activation energy is found, the conclusion is sometimes drawn that the reaction will proceed slowly or with difficulty. This is an incorrect assumption, because the reaction may actually proceed quite fast, namely at high temperature. The point is

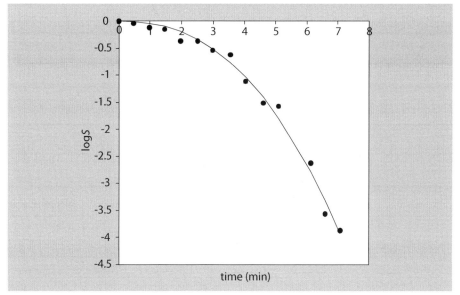

Figure 4.17. Fit of the Weibull model to the inactivation of Salmonella typimurium. Weibull parameters α = 2.4 min and β = 2.4. Adapted from Mackey and Derrick (1986).

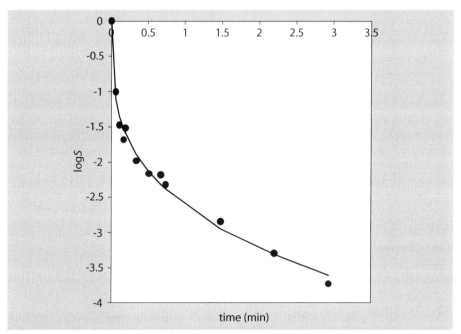

Figure 4.18. Fit of the Weibull model to the inactivation of Salmonella enteriditis *in egg yolk. Weibull parameters* $\alpha = 0.003$ *min and* $\beta = 0.3$. *Adapted from Michalski et al. (1999).*

that a high activation energy indicates a strong temperature dependence, that is to say it will run very slowly at low temperatures, but very fast at high temperatures. What is relevant for foods is that chemical reactions (e.g. the Maillard reaction) have a 'normal' activation energy of about 100 kJ.mol^{-1}, whereas the inactivation of micro-organisms can be characterized by a high activation energy, say, 300 kJ.mol^{-1} (even though, as already mentioned, it is incorrect to express it in this way because the killing of micro-organisms is not a simple elementary reaction). Figure 4.19 illustrates this difference in temperature sensitivity. These phenomena are exploited in processes such as HTST (high-temperature short-time heating) and UHT (ultra high-temperature treatment). These processes are designed by choosing time-temperature combinations such that desired changes are achieved (microbial inactivation), whilst undesired changes (chemical reactions leading to quality loss) are minimized. Another important consequence for foods is that reactions with relatively low activation energy will continue at a measurable rate at low temperatures, for instance, during storage, leading to a limited shelf life.

Several types of reaction can occur in foods, as discussed in section 4.2. Chemical reactions more or less obey the Arrhenius' equation as would be expected. Several physical reactions are less temperature dependent and often diffusion controlled. The concept of activation energy does not actually apply to physical reactions (such as coalescence, aggregation) because there are no molecular rearrangements.

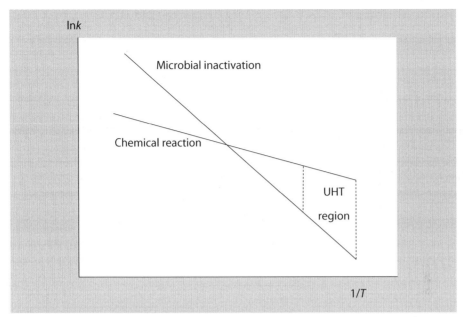

Figure 4.19. Schematic presentation of the temperature dependence of a chemical reaction and microbial inactivation. The UHT region is characterized by time-temperature combinations that induce enough microbial inactivation and limited chemical reactions.

However, physical phenomena do usually have an energy barrier (due to, for instance, electrostatic repulsion) which provide stability to colloidal systems. Hence, the concept of a kind of activation energy does apply but not with a temperature dependence as in the case of chemical reactions. The effect of temperature will be mainly on the rate of encounters. Sometimes, activation energies are reported for physical phenomena such as the temperature dependence of diffusion or viscosity. This would seem to be impossible, since there is nothing to activate and there is no reaction. As discussed before, the point is that the temperature dependence of diffusion, for example, apparently obeys Arrhenius' law in several systems, but the parameter that comes out of it does not have the physical meaning of an activation energy! The same is true for models that describe temperature dependence of viscosity in foods undergoing a glass transition. The Williams-Landel-Ferry (WLF) model is often used to model such dependence. It is reported that the Arrhenius' model does not function well for such cases. The WLF model was developed for polymers but is nowadays also applied to foods that undergo glass transitions. The WLF model is actually not comparable to the Arrhenius' equation, because it does not consider activation energies; it is empirical in nature and attempts to model viscosity as a function of temperature. The WLF model is:

$$\ln\left(\frac{\eta_v}{\eta_{v.g}}\right) = \frac{-C_1(T'-T'_g)}{C_2+(T'-T'_g)} \tag{4.33}$$

C_1 (dimensionless) and C_2 (°C) are parameters of the WLF model, T'_g is the glass transition temperature (°C) and $\eta_{v,g}$ is the viscosity at the glass transition temperature.

Quite different results are obtained with protein denaturation and microbial inactivation. (Microbial inactivation is according to some authors due to enzyme, i.e. protein, denaturation. It is questionable whether this is the sole cause of inactivation.) Protein denaturation is characterized by a high activation energy and this is compensated for by a high pre-exponential factor. As a result, the temperature dependence of such reactions is very high, much higher than that of chemical reactions.

With biochemical reactions, i.e. enzyme-catalysed reactions, there is moderate temperature dependence, as is to be expected for catalysed reactions. It is of interest to note that the rate enhancement by enzymes, as compared to the uncatalysed reaction, is much higher at lower temperatures than at higher temperatures: Figure 4.20 gives a schematic impression. Enzymes lower the activation energy considerably (as compared to the uncatalysed reaction), whereas the pre-exponential factor is only changed a little. However, with enzyme-catalysed reactions, enzymes become inactivated above a certain temperature, and the catalysed reaction effectively comes to an end. Most enzymes relevant in food tend to become inactivated between 50-80 °C, though some notably heat-resistant enzymes are known. The same goes for microbial growth: first there is an increase with temperature but eventually microbes start to die. A highly schematic picture of the effect of temperature on microbial growth and enzyme action alike is shown in Figure 4.21; it should be noted that

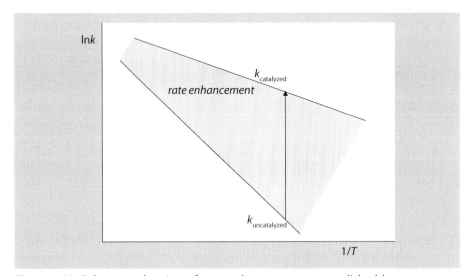

Figure 4.20. Schematic drawing of rate enhancement accomplished by enzymes as compared to the uncatalysed reaction.

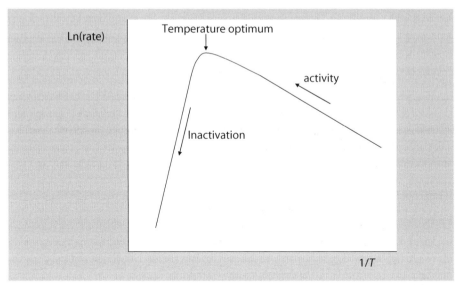

Figure 4.21. Schematic presentation of the effect of temperature on growth and activity of enzymes and micro-organisms and their inactivation.

the actual response to temperature can be time-dependent. In the case of micro-organisms, there is also a minimum temperature below which there is no growth. For that reason, empirical relations have been derived to describe temperature dependence of microbial growth, for instance, the square root model that describes the effect of temperature on the maximum growth rate:

$$\sqrt{\mu_{max}} = b_1[T - T_{min}][1 - \exp(c_1(T - T_{max}))] \tag{4.34}$$

In this equation, b_1 and c_1 are fit constants, and $T_{min,max}$ are the minimum and maximum temperature for growth of the micro-organism under study. In the same way, empirical relations are described for temperature dependencies of lag time λ. See McKellar and Lu (2004) for more details.

Photochemical reactions and radical reactions are not or are only slightly temperature dependent because the changes at the molecular level hardly depend on thermal energy. Both types of reactions are of importance in foods. Photochemical reactions cause, for instance, oxidation of vitamins, they may activate certain enzymes, and they may cause flavour defects. Radical reactions are most notable for oxidation reactions (of unsaturated fats or of vitamins).

4.5.2. Shelf life

Once foods are designed and manufactured, there may be a considerable lapse of time, because of distribution and storage, before the food is actually consumed.

Ideally, one would like the quality not to deteriorate. However, while some foods gain in quality during storage (e.g. cheese, wine), most foods will loose their quality gradually over time. Once again, we need kinetics to make some sensible predictions. Firstly, micro-organisms should not be given the chance to grow. For sterilized foods, this is not a problem, but otherwise it really depends on the type of food and on the environmental temperature during storage. Growth models are very helpful in this respect to be able to predict a 'use before date' label. The most practical problem here is that temperature may fluctuate considerably during storage, which can have a detrimental effect on microbial quality; see for instance, Labuza (1979) and Labuza and Fu (1993).

If microbial deterioration is under control, the next problem is most likely biochemical and chemical degradation. Biochemical changes can be prevented by the inactivation of enzymes, but if foods are minimally processed, residual enzymatic activity can be a problem. A striking example of such a problem was in the early days of UHT processing of milk when very heat-stable proteases and lipases caused lipolysis and proteolysis, respectively, resulting in unacceptable milk after some 6 weeks of storage at ambient temperature, even though microbial stability was not a problem. This problem can be solved by either giving a slightly more intense heat treatment than strictly needed for microbial stability, or by cooling the product, not for microbial reasons but such that the enzymatic reaction is slowed down. Chemical reactions will always continue, at low temperatures too, but their rate will, of course, slow down substantially at low temperatures. Again, models are very helpful for making predictions for a 'best before date'. Temperature has, of course, a significant effect on shelf life. The Arrhenius' equation is very helpful to make predictions for chemical and biochemical changes; if temperature fluctuates, this should of course be taken into account by using variable temperature kinetics instead of isothermal kinetics. Physical changes are not so strongly temperature dependent, though there may be indirect effects, such as physical reactions due to chemical changes. For instance, sedimentation may become possible after enzymatic degradation of cell wall polysaccharides. Nowadays, the use of models is very helpful to make sensible predictions regarding shelf life. Some specific references related to shelf-life modelling are Man (2002), Kilcast and Subramaniam (2000), Eskin and Robinson (2001).

Shelf life in relation to sensorial aspects is of course very important, and this cannot be so easily modelled by chemical models. Sensorial tests are needed to determine when products become unacceptable for consumers. Based on such data, probabilistic models can be built. An example of this is the use of the Weibull model, originally developed to model failure times in electronic and mechanical devices, but also applicable to food. Such a model will give a prediction in the sense that after a specified time a certain percentage of the consumers will reject a product. Some useful references for this type of shelf-life modelling are: Gacula (1975), Gacula and Kubala (1975), Thiemig *et al.* (1998), Al Khadamany *et al.* (2002, 2003), Hough *et al.* (1999), Hough (2010).

4.5.3. Design of foods via modelling

The above models represent only a minority of the many models that can be found in literature in relation to food quality. As mentioned before, models should be applied with care, that is to say the user should be aware of the possibilities and limitations. In this section, we would like to spend a little time explaining how to use models in designing foods. Engineering design requires modelling to complete the design task, which can be described as a set of technical activities within a product development process (Otto and Wood, 2001). Modelling is based on an understanding of relevant physical phenomena.

Suppose we are asked to develop a new pizza that will not be frozen but will be supplied through a retailer in a cooled chain. The question then is how models can help to design such a product. The first thing to do is make an inventory of possible quality issues, as in Table 4.6.

The problem with such a pizza, as compared to a pizza that is prepared at home and immediately consumed, is that all kinds of chemical, biochemical and physical changes take place during the relatively long time that such a semi-prepared product spends in the food chain. Although Table 4.6 is not an exhaustive list of all possible quality issues, it does show that many aspects have to be taken into account. Once the most relevant quality issues have been tackled, suitable models need to be found. Furthermore, these models must be coupled to determine at which stage the various quality attributes are at stake, and a management decision will need to be taken with respect to what is acceptable and what not anymore. Models can be used here to support the decision-making process. It may be that one measure to prevent one process has a great effect on another quality issue. For instance, introducing a heating step to the vegetables before assembling the pizza to reduce the number of micro-organisms may have a detrimental effect on the texture and colour of the vegetable toppings if the heating is too intense. Models can then be used to find out the best combinations of treatments to reach the optimal quality. Obviously, models can also be used to predict shelf life, and it may well be that the shelf life is not determined by microbial limits but by physical phenomena. Of course, appropriate technological measures can also be taken, such as modified atmosphere packaging, to prevent microbial spoilage but also to prevent oxidation and enzymatic activity. Models can also be used in conjunction with the parameters applied in a certain technology.

It is hoped that with the examples of models given earlier in this chapter and the approach used in this section, the reader is convinced that modelling and food science are a necessary and useful combination in the food design process.

Table 4.6. Possible quality issues in the design of a non-frozen pizza with a vegetable topping.

Quality issue	Causes	Models needed	Input needed
Spoilage by micro-organisms	Microbial growth	Bacterial growth models	Relevant contaminating micro-organisms, temperature profiles in the chain
Possible technological actions			
Heating	Killing of microbes	Inactivation models	Kinetic parameters
Modified atmosphere packaging	Prevention of growth		
Loss of nutritive value	Oxidation	Models describing oxidation	Effect of oxygen, rate constants
Possible technological actions			
Vacuum packaging	Removal of a reactant	Gas permeation models	Permeability data of packaging material
Texture loss of vegetables on topping	Enzymatic and chemical degradation of cell walls	Models describing texture loss	Characteristics of relevant enzymes (denaturation temperature, v_{max}, k_m)
Possible technological actions			
Blanching	Inactivation of enzymes	Enzyme inactivation models	Kinetic parameters
Loss of colour of vegetables	Chlorophyll degradation	Chemical models describing chlorophyll loss	Temperature profiles, ph
	Enzymatic browning	Enzymatic models for enzymatic browning	Characteristics of relevant enzymes (denaturation temperature, v_{max}, k_m)
Possible technological actions			
Modified atmosphere packaging	Removal of oxygen		
Blanching	Inactivation of enzymes	Enzyme inactivation models	Kinetic parameters
Loss of taste and flavour	Partitioning, diffusion	Physical models describing partitioning phenomena, diffusion models	Partitioning coefficients, activity coefficients
Possible technological actions			
Barrier coating	Prevention of diffusion	Diffusion and partitioning models	Diffusion and partition coefficients
Modified atmosphere packaging	Removal of reactants		Oxygen and water permeability of packaging material
Blanching	Inactivation of enzymes	Enzyme inactivation models	Kinetic parameters

Table 4.6. Continued.

Quality issue	Causes	Models needed	Input needed
Texture of dough	Water diffusion from vegetables to dough	Physical model describing diffusion of water	Water activity values, effective diffusion coefficients
	Starch retrogradation, protein aggregation	Physical models describing protein aggregation, starch retrogradation	Diffusion coefficients, parameters describing retrogradation
Possible technological actions			
Barrier coating	Prevention of water migration	Diffusion models	Water activity, sorption isotherms

4.6. Statistical design of experiments in relation to food product design

The discussion above was about modelling specific quality attributes. One of the difficulties of food product and process design is that there is a wide range of materials that can be used as raw material and/or ingredients, as well as a range of processing possibilities. To reduce this wide range in a sensible way, experiments have to be set up that help in choosing the right conditions, and this is another important point at which statistics come in. Conventionally, this is done by the so-called method of 'one factor at a time', i.e. varying one factor while holding other factors constant, finding the optimal value of that factor, fixing it at that level and then varying another factor. It is basically a method of trial and error. In doing so, one neglects the possibility of interaction between the various factors studied and as a result the design will be suboptimal. Statistical techniques are available to design experiments in such a way that these interactions can be found. Software to do this is now widely available, also for the specific area of food design (a search on the Internet using the keywords 'design of experiments' and 'foods' will give many hits). This is not the place to discuss this in depth, so a few remarks will suffice. A very nice introduction to the topic is given by Hu (1999). Basically so-called factorial designs are used; each input variable is set at

Table 4.7. Example of a 2^2 factorial design. X_1 and X_2 are the factors to be varied at a low (-1) and a high (+1) level.

Exp. No.	X_1	X_2
1	-1	-1
2	-1	+1
3	+1	-1
4	+1	+1

a high and a low level, and combinations for all input variables are tested. Figure 4.22 gives a schematic picture of a 2^2 factorial design in comparison to a one-factor-at-a-time design. In the latter case, one factor is fixed at a certain level, and the other one is varied, for instance, varying cooking time while holding the pH constant, and vice versa, varying the pH while cooking time is kept constant. In the factorial design the factors are varied systematically at a high and a low level (see Table 4.7).

This allows interaction effects to be detected, and is also much more efficient in terms of time and money because usually less experiments have to be done to obtain the same information (see also Figure 4.22). There are actually quite a few designs that can be used: full factorial designs, fractional factorial design, central composite designs, Taguchi designs. The interested reader is referred to Arteaga *et al.* (1994), Hu (1999) and Roy (2001) for further information.

Of course, one needs to have knowledge of the system under investigation in order to choose the right factors as well as their levels. For that reason, the discussion given in the previous section allows us to make sensible estimates of factors and ranges. Systematic experimental design will also lead to models, but these are usually statistical models in the form of polynomials, not kinetic models. For instance, for two factors a model could look like:

$$R = \beta_0 + \beta_1 X_1 + \beta_2 X_2 + \beta_{12} X_1 X_2 \tag{4.35}$$

in which R is the response one is looking for (for instance, a certain texture, or the inactivation of an enzyme), β_0 represents the average, β_1 and β_2 take the linear effect

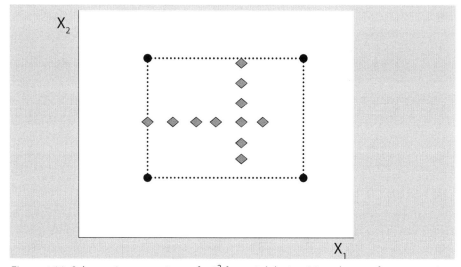

Figure 4.22. Schematic comparison of a 2^2 factorial design (•) and a one-factor-at a time design (♦). X_1 and X_2 are the factors to be varied.

of factor X_1 and X_2 into account, and β_{12} the interaction effects of X_1 and X_2. Such models can also contain quadratic terms. In any case, it should be clear that such systematic approaches allow for the extraction of the greatest amount of meaningful information from the fewest number of experiments and at the lowest costs.

4.7. Future developments

A development that could be useful for the aim of food product design is unconventional modelling techniques such as neural networks, fuzzy logic and Bayesian belief networks (Corney, 2000; Van Boekel *et al.*, 2004; Phan *et al.*, 2010). These modelling techniques come from the area of artificial intelligence and do not yet have many food applications, but they will probably become more important in the future. The characteristics of neural networks and fuzzy logic are that such models can learn when they process data. Bayesian belief networks can deal with uncertainty (via probability distributions) and they allow the use of expert knowledge (Van Boekel *et al.*, 2004). Hence, they are useful for decision support systems.

As regards food processing, modelling techniques are very promising with respect to computational fluid dynamics (CFD). Although this is an application area that requires heavy computing, it allows the engineer to do calculations that were not hitherto possible. Applications are in the area of heat and mass transfer. It allows a fine-tuning of processing in terms of heating, for instance. An overview can be found in Nicolaï *et al.* (2001).

References

Al Kadamany, E., Toufeili, I., Khattar, M., Abou-Jawdeh, Y., Harakeh, S. and Haddad, T. (2002). Determination of shelf life of concentrated yogurt (Labneh) produced by in-bag straining of set yogurt using hazard analysis. Journal of Dairy Science 85:1023-1030.

Al Kadamany, E., Khattar, M., Haddad, T. and Toufeili, I. (2003). Estimation of shelf-life of concentrated yogurt by monitoring selected microbiological and physico-chemical changes during storage. Lebensmittelwisschenschaft und Technologie 36:407-414.

Anema, S.G., Lowe, E.K. and Li YuMing (2004). Effect of pH on the viscosity of heated reconstituted skim milk. Intern. Dairy J. 14: 541-548.

Anthon, G.E., Sekine, Y., Watanabe, N. and Barrett, D.M. (2002). Thermal Inactivation of Pectin Methylesterase, Polygalacturonase, and Peroxidase in Tomato Juice. J. Agric. Food Chem. 50:6153-6159.

Arteaga, G.E., Li-Chan, E., Vazquez-Arteaga and Nakai, S. (1994). Systematic experimental designs for product formula optimisation. Trends Food Sci. Technol. 5: 243-254.

Audu, T.O.K., Loncin, M. and Weisser, H. (1978). Sorption isotherms of sugars. Lebensmittelwisschenschaft und Technologie 11:31-34.

Blandamer, M.J., Engberts, J.B.F.N., Gleeson, P.T. and Reis, J.C.R. (2005). Activity of water in aqueous systems; a frequently negelected property. Chemical Society Reviews 34:440-458.

Box, G.E.P. (1976). Science and statistics. Journal of American Stat. Assoc 71:791-799.

Brands, C.M.J. and Van Boekel, M.A.J.S (2002). Kinetic modelling of reactions in heated monosaccharide-casein systems. J. Agric. Food Chem. 50: 6725-6739.

Brul, S., Van Gerwen, S. and Zwietering, M. (eds.) (2007) Modelling microorganisms in food. Woodhead Publishing Ltd., Cambridge, UK.

Burnham, K.P. and Anderson, D.R..(1998). Model Selection and Inference. A practical information and-theoretic approach. New York, Springer Verlag.

Chase, A.M., Meier, H.C and Menna, V.J. (1962). The non-competitive inhibition and irreversible inactivation of yeast invertase by urea. J Cellular and Comparative Physiology 59:1-13.

Corney, D. (2000). Designing food with Bayesian Belief Networks. In: Parmee I (ed.) Adaptive computing in design and manufacture, pp. 83-94, University of Plymouth, UK.

Damodaran, S., Parkin, K. and Fennema, O.R. (eds.) (2009) Fennema's Food Chemistry. 4th edition. CRC/Taylor & Francis, Boca Raton, FL, USA.

Eskin, N.A.M. and Robinson, D.S. (2001). Food Shelf Life Stability. Chemical, Biochemical and Microbiological changes. Boca Raton, USA, CRC Press.

Fennema, O.R. (1996). Food Chemistry, 3rd edition. Marcel Dekker, New York, NY, USA.

Gacula, M.C. (1975). The design of experiments for shelf life study. Journal of Food Science 40:399-403.

Gacula, M.C. and Kubala, J.J. (1975). Statistical models for shelf life failures. J Food Sci 40:404-409.

Gibson, A.M., Bratchell, N. and Roberts, T.A. (1988) Predicting microbial growth: growth responses of salmonellae in a laboratory medium as affected by pH, sodium chloride, and storage temperature. Int J Food Microbiol 6:155-178.

Hayes, M.G., Hurley M.J., Larsen, L.B., Heegaard C.W., Magboul, A.A.A., Oliveira, J.C., McSweeney, P.H. and Kelly A.L. (2001). Thermal inactivation kinetics of bovine cathepsin D. J. Dairy Res. 68:267-276.

Horak, F.P. (1980). Über die Reaktionskinetik der Sporenabtötung und chemischer Veränderungen bei der thermischen Haltbarmachung von Milch zur Optimieruung von Erhitzungsverfahren. PhD thesis Technical University of Munich, Germany.

Hough, G. (2010). Sensory shelf life estimation of food products. CRC/Taylor & Francis, Boca Raton, FL, USA.

Hough, G., Puglieso, M.L., Sanchez, R. and Mendes da Silva, O. (1999). Sensory and microbiological shelf life of a commercial ricotta cheese. Journal of Dairy Science 82:454-459.

Hu, R. (1999). Food Product Design. A computer-aided statistical approach. Technomic Publishing Company.

Jay, J.M., Loessner, M.J. and Golden, D.A. (2005). Modern Food Microbiology. 7th Edition. Springer, New York, NY, USA.

Kilcast, D. and Subramaniam, P. (2000). The stability and shelf-life of food. Cambridge, Woodhead Publishing Ltd. and CRC Press LLC.

Labuza, T.P.(1979). A theoretical comparison of losses in foods under fluctuating temperature sequences. Journal of Food Science 44:1162-1168.

Labuza, T.P. (1983). Reaction kinetics and accelerated tests. Simulation as a function of temperature. In: I. Saguy (ed.) Computer Aided Techniques in Food Technology. Marcel Dekker, New York, NY, USA, p. 71-115.

Labuza, T.P. and Fu, B. (1993). Growth Kinetics for Shelf-Life Prediction: Theory and Practice. Journal of Industrial Microbiology 12(3/5):309-323.

Le Meste M., Champion, D., Roudaut, G., Blond, G. and Simatos, D. (2002). Glass transition and Food Technology: a critical appraisal. Journal of Food Science 67:2444-2458.

Levine, L. (1997). Food process development and scale up using modelling and simulation. Chemistry & Industry 9:346-349.

Lievonen, S.M., Laaksonen, T.J. and Roos, Y.H. (1998). Glass transition and reaction rates: nonenzymatic browning in glassy and liquid systems. J Agric Food Chem 46:2778-2784.

Man, D. (2002). Shelf life. Oxford, UK, Blackwell Science Ltd.

Mackey, B.M. and Derrick C.M. (1986). Elevation of the heat resistance of Salmonella typhimurium by sublethal heat shock. J Appl Bacteriol 61:389-39.

Manski, J.M., Kretzers, I.M.J., Van Breknk, S., Van der Goot, A.J. and Boom, R.M.(2007). Influence of dispersed particles on small and large deformation properties of concentreated caseinate composites. Food Hydrocolloids 21: 73-84.

Marangoni, A.G. (2003). Enzyme Kinetics. A modern approach. Hoboken NY, USA, Wiley Interscience

Martins, S.I. and Van Boekel, M.A. (2003). Kinetic modelling of Amadori N-(1-deoxy-D-fructos-1-yl)-glycine degradation pathways. Part II-kinetic analysis, Carbohydr Res. 338(16): 1665-78.

Martins, S.I.F.S. and Van Boekel, M.A.J.S. (2005a). A kinetic model for the glucose/glycine Maillard reaction pathways. Food Chemistry 90:257-269.

Martins, S.I.F.S. and Van Boekel, M.A.J.S. (2005b). Kinetics of the glucose/glycine Maillard reaction pathways: influences of pH and reactant initial concentrations. Food Chemistry 92:437-448.

McKellar, R.C. and Lu, X. (eds) (2004). Modeling microbial responses in food, pp. 1-20. CRC Press, Boca Raton.

Michalski, C.R., Brackett R.E., Hung, Y.-C. and Ezeike, G.O.I. (1999). Use of capillary tubes and plate heat exchanger to validate U.S. Department of Agriculture pasteurization protocols for elimination of Salmonella enteritidis from liquid egg products. J Food Prot 62:112-11.

Mínguez-Mosquera, M.I. and Gandul-Rojas, B. (1994). Mechanism and kinetics of carotenoid degradation during the processing of green table olives. J. Agric.Food Sci. 42:1551-1554.

Nicolaï, B.M., Verboven, P. and Scheerlinck, N. (2001). The modelling of heat and mass transfer. In: Food Process Modelling. L.M.M. Tijskens, M.L.A.T.M. Hertog, B.M. Nicolaï (eds.). Woodhead Publishing Ltd., pp. 60-86.

Otto K. and Wood K. (2001). Product Design: Techniques in Reverse Engineering and new product development. London, Prentice Hall International.

Owusu-Apenten R.K. (2005). Introduction to Food Chemistry. Boca Raton, CRC Press.

Peleg, M. and Cole, M.B. (1998). Reinterpretation of microbial survival curves. Crit Rev Food Sci Nutr 38:353-380.

Phan, V.A., Van Boekel, M.A.J.S., Dekker, M. and Garczarek, U. (2010) Bayesian networks for food science: theoretical background and potential applications. In: S.R. Jaeger and H. MacFie (eds.) Consumer-driven innovation in food and personal care products. Woodhead Publishing Ltd., Cambridge, UK, pp. 488-513.

Roos, Y. (1995). Characterization of Food Polymers Using State Diagrams. Journal of Food Engineering 24:339-360.

Roos, Y.H., Karel, M. and Kokini, J.L. (1996) Glass transitions in low-moisture and frozen foods: Effects on shelf life and quality. Food Technology 50(11):95-108.

Roy, R.J. (2001). Design of experiments using the Taguchi approach. Wiley-Interscience, New York, 538 pp.

Saguy, I., Kopelman, IJ. and Mizrahi, S. (1978). Thermal kinetic degradation of betanin and betalamic acid. J Agric Food Chem 26: 360-362.

Thiemig, F., Buhr, H. and Wolf, G. (1998). Charakterisierung der Haltbarkeit und des Verderbsverhaltens frischer Lebensmittel. Fleischwirtschaft 78:152-154.

Van Boekel, M.A.J.S. (1996). Statistical aspects of kinetic modeling for food science problems. J Food Sci 61:477-485, 489.

Van Boekel, M.A.J.S. (1999). Testing of kinetic models: usefulness of the multiresponse approach as applied to chlorophyll degradation in foods. Food Res Int 32:261-269.

Van Boeke,l M.A.J.S. (2000). Kinetic modelling in food science: a case study on chlorophyll degradation in olives. J Sci Food Agric 80:3-9.

Van Boekel, M.A.J.S. (2002). On the use of the Weibull model to describe thermal inactivation of microbial vegetative cells. Int J Food Microbiol 74:139-159.

Van Boekel, M.A.J.S. (2008). Kinetic modeling of reactions in foods. CRC/Taylor & Francis, Boca Raton, FL, USA.

Van Boekel, M.A.J.S., Stein, A. and Van Bruggen, A. (2004). Bayesian Statistics and Quality Modelling in the Agro Food Production Chain. Kluwer Academic Press, Dordrecht, the Netherlands.

Walstra, P. (2003). Physical chemistry of foods. Marcel Dekker, New York, NY, USA.

Whitaker, J.R., Voragen, A.G.J. and Wong, D.W.S. (2003). Handbook of Food Enzymology. Marcel Dekker, New York, NY, USA.

5. Barrier technology in food products

Remko Boom[1], Karin Schroën[1] and Marian Vermuë[2]
[1]Laboratory of Food Process Engineering, Wageningen University [2]Bioprocess Engineering Group, Wageningen University

5.1. Introduction

5.1.1. Edible coatings and barriers in food applications

Modern food products become more and more complex. This is due to a variety of trends in the consumer's wishes. There is, for example, a clear wish for products that are minimally processed but at the same time remain fresh up to the moment of consumption. There is also a desire for products that can either be prepared quickly or are ready-to-eat as they are offered. In many cases products consist not only of many ingredients, but in fact of incompatible ingredients. Therefore, barrier technology is increasingly applied to be able to combine these components in one product. In nature, this is for example done through compartmentalisation in a cell. Some examples of typical food products with incorporated barriers are given below to illustrate the importance of barrier technology.

The primary protection of many products is the epidermis that protects it from influences from outside. However, as soon as this layer is damaged, e.g. during food processing, the food needs to be protected in another way. The oldest form of barrier technology is packaging, used to protect the product from contamination or damage from outside, to keep oxygen out, and possibly maintain a protective atmosphere around the product. Apples have a natural waxy coating (cuticle) that prevents evaporation and drying out of the fruit. The use of pesticides in modern agriculture necessitates washing the fruit before it can be distributed. Mechanical harvesting causes damage to the protective coating. Therefore, an artificial coating is applied to avoid quality deterioration (Chen, 1995). Baked products may have a tendency to stick to the wrapping. Therefore a coating may be applied that prevents the product from sticking. A cookie may have a filling with a fruit jelly. Direct application of the jelly to the cookie results in migration of water from the jelly to the cracker, which then loses its glassy state and becomes soft. A coating may be applied to prevent the migration of moisture, or at least postpone it sufficiently long (Kim, 1998). A fruit product may be susceptible to microbial contamination, but may need respiration. Application of a porous coating that is impermeable to micro-organisms, but that permits the transport of gases, can prolong the shelf life of the product. The same applies to cheeses, which can be infected with fungi that may spoil the cheese during or after ripening. Application of an edible coating, instead of traditional packaging applied directly to the surface of the product, may reduce the total waste production considerably, that is, if the integrity of the coating remains in tact. It is estimated that around the turn of the century, food packaging generated approx. 32 million tons of waste in the United States alone (Chen, 1995). A small reduction in the amount of packaging material, through the application of edible coatings from renewable resources, may therefore result in a significant reduction in the amount of waste generated. This list of the applications of barrier technology is by no means complete; it is in fact growing rapidly.

5.1.2. Contents of this chapter

During the design process of more complex food products, the use of a barrier may be considered. The selection of an applicable coating is no trivial matter, however. Many different materials are available, and many different forms of barriers are possible (Table 5.1). The design of a barrier that complies with the requirements on product quality, shelf life and robustness for handling is an intrinsic part of the total product design process.

In this chapter, we will give some basics on the underlying theory for the design of an edible barrier or coating on structured food products. The data given are primarily linked to the water barrier function. However, the theory is perfectly applicable to any other type of permeant as well, such as oxygen.

The basis of the design of an appropriate barrier lies in thermodynamics (chemical potential differences, sorption isotherms and driving forces) and in dynamics (mobility and diffusivity). These two domains will be the basis on which the guidelines for the design of barriers in food, as presented in this chapter, are founded.

Table 5.1. Some uses of edible films and coatings (Greener Donhowe, 1994).

Use	Appropriate types of films
Retardation of moisture migration	Lipid[1], heterogeneous[2]
Retardation of gas migration	Biopolymers[3], lipid or heterogeneous
Retardation of migration of oils or fats	Biopolymers
Retardation of solutes	Biopolymers, lipid or heterogeneous
Improvement of structural integrity or handling properties	Biopolymers, lipid or heterogeneous
Retain volatile flavour components	Biopolymers, lipid or heterogeneous
Convey food additives	Biopolymers, lipid or heterogeneous

[1]Lipids may be waxes, acyl-glycerols, and fatty acids.
[2]Heterogeneous implies the use of more than one phase, usually a lipid and an aqueous, bio-polymeric phase, applied as emulsions, or as multi-layered films.
[3]Biopolymers include solutions of proteins, cellulose derivatives, alginates, pectin, starch or other polysaccharides.

5.2. Thermodynamics and sorption

5.2.1. The thermodynamic potential and the activity

A barrier is designed to prevent or hinder the transfer of one or more components. These components apparently are subject to a driving force to permeate the barrier and may enter another part of the food product. The field of thermodynamics quantifies this driving force. It can be expressed as a potential, or as an activity. The relation of the thermodynamic potential μ_i of component i and its activity a_i is given by:

$$\mu_i = \mu_i^\circ\left(p_i^{sat}, T\right) + RT \ln\left(a_i\right)$$ (5.1)

in which μ_i^0 is a reference that is not dependent on the composition (but is dependent on pressure and temperature). The reference temperature and pressure used for μ_i^0 can be chosen freely. In this case, the saturated vapour pressure is chosen. Under these conditions the activity in a pure liquid is equal to the activity in the saturated vapour pressure above this liquid under equilibrium conditions.

In Equation 5.1, a_i denotes the activity of a component, and its value depends on the aggregation state of the system (liquid, solid or gaseous); see also Chapter 3 on kinetic modelling.

Gases

For gases, the activity is generally equal to the partial vapour pressure p_i, divided by the saturated vapour pressure of component i, p_i^{sat} at those conditions: $a_i = p_i/p_i^{sat}$. The value of p_i^{sat} is equal to the pressure that would exist in a gas phase containing *only* component i, above a liquid phase of *only* component i. The value of p_i^{sat} is dependent on temperature (it increases quickly with temperature), and is characteristic for each component. For example, the saturation pressure of water is 2.3 kPa at 20 °C, 19.7 at 60 °C, and 100 kPa at 100 °C (which means that at 100 °C, the saturation pressure of water is equal to the ambient pressure, and thus liquid water at 100 °C has a vapour above it containing 100% water molecules: it boils).

The most frequently used method for finding the values of the saturation pressure, is with the help of the empirical *Antoine's equation*:

$$\log p_i^{sat} = A - \frac{B}{T+C}$$ (5.2)

The constants A, B and C are properties of the component, and are tabulated in many textbooks and reference books. The largest compendium of data in this area is to be found in Gmehling and Onken (1977). For water vapour, its values are: $A = 8.07131$; $B = 1730.630$ and $C = 233.426$, valid for values of T between 1 and 100 °C, and T expressed in °C. Please note, that the resulting p_i^{sat} is in mmHg (1 atmosphere = 101.325 kPa = 760 mmHg).

Liquids and non-crystalline polymeric systems

In liquids, the value of the activity is not directly given. The quantity a_i is given by:

$$a_i = \gamma_i x_i \tag{5.3}$$

in which, γ_i is the (*rational*) *activity coefficient*, and x_i is the *mole fraction* of component *i*. Please note, the activity coefficient used in Chapter 3 is different from the activity coefficient mentioned here that uses molar fractions; we use this one because it is practical when working with water activity. In ideal systems, the activity coefficient is one, and the activity is equal to the mole fraction, x_i. This is the case in many dilute solutions. The value of the activity coefficient can be quite different from one (higher and lower than unity) for more concentrated solutions. Ionic components in general behave non-ideally, even at moderate to low concentration. Their behaviour is difficult to describe with simple universal models; an option would be the extended Debije-Huckel, mean spherical approximation, but it would lead too far to explain this theory here. The interested reader is referred to the textbook of Tiny van Boekel on Kinetic Modeling of Reactions in Foods, Chapter 5 (2008).

The behaviour of molecules in a solution is something that has been studied for many years (in fact, centuries), but is still not well understood. Therefore, there is no clear-cut theory for the description of the values of the activity coefficient of a given solution. There are some rough guidelines. In general, a very dilute solution will behave ideally; the concentration above which the behaviour is no longer ideal varies. A solution of coffee with sugar will usually behave rather ideally, unless you really make syrup out of it. But an apolar molecule, e.g. a fat molecule, or hexane, in water, will already have a high activity coefficient at rather low concentrations (although the activity coefficient of the solvent itself *always* goes to unity at sufficiently low concentrations of the solute). You can deduce this from the fact that it only partially dissolves in water.

Further, solutions in which only Van der Waals type interactions take place, tend to behave ideally. Van der Waals may be described as the type of interactions that remain when all other interactions like ionic interactions, dipole-dipole, dipole-induced dipole and hydrogen or nitrogen bonds are absent due to the nature of the two phases. Aqueous solutions, with many hydrogen bonds, generally do not behave ideally, unless they are dilute. Unfortunately, many food products need to be regarded as more or less concentrated aqueous solutions, which behave non-ideally.

Many semi-empirical models are available for calculating the activity coefficient of a given solution: models named after their developer, such as Wilson and Van Laar, or models with names that give a clue to their basis, such as the Non-Random Two-Liquid model. There is no existing model that predicts reality completely, so you have to see which model gives good results in a certain situation.

The total driving force for a component to move (e.g. through a barrier) is equal to the negative gradient of the thermodynamic potential, with respect to the location:

$$F_i = -\frac{d\mu_i}{dz} \qquad (5.4)$$

The unit of the potential μ_i is J/mole; since the unit of the spatial coordinate z is meters, the unit of F_i is in Newton per mole: it is the force that acts on 1 mole of molecules of permeant. The larger the driving force caused by this gradient, the larger the rate of mass transfer. For almost all practical systems, it has been found that the mass transfer is proportional to the size of the driving force (see also the next section).

The driving force for mass transfer can be approximated by taking a difference approach:

$$F_i = -\frac{\Delta\mu_i}{\Delta z} \qquad (5.5)$$

Let us consider mass transfer of a component i from a food matrix (phase A) through a thin layer of a barrier material (phase B). The chemical potential at the food matrix and the barrier matrix will be different because both phases have different properties.

$$\mu_i^A = \mu_i^\circ + RT \ln\left(a_i^A\right) \text{ and } \mu_i^B = \mu_i^\circ + RT \ln\left(a_i^B\right) \qquad (5.6)$$

If there is no mass transfer between the food matrix and a barrier material the chemical potentials should be equal in both phases: $\mu_i^A - \mu_i^B$. The phases are in *equilibrium,* which indicates that the activity of the permeant in the food matrix should be equal to the activity in the barrier material.

$$a_i^A = a_i^B \quad \rightarrow \quad \gamma_i^A x_i^A = \gamma_i^B x_i^B \qquad (5.7)$$

Be aware that the activity coefficient γ_i will not necessarily be the same at both sides. In general, one chooses a barrier material that does not dissolve the permeant as well as the matrix that is to be protected – for example, to protect a product from desiccation, one may use a hydrophobic material. For such a material, typical values of activity coefficients are quite high; much higher than the value in the food matrix. Equation 5.7 shows that the mole fraction of the permeant in the barrier material in this case will be much lower than the mole fraction in the food matrix (at equilibrium).

5.2.2. Sorption isotherms

Often, the tendency of a material to dissolve a permeant is not expressed in terms of an activity coefficient, but in terms of a sorption isotherm: this usually relates the composition of one phase to the composition in the other (at a specific temperature).

In fact, the equation given in the former section is a specific form of a sorption isotherm:

$$a_i^A = a_i^B \quad \rightarrow \quad x_i^B = \left(\frac{1}{\gamma_i^B}\right)a_i^A = m\,a_i^A \tag{5.8}$$

This isotherm is called the Henry relation and the parameter m is called the Henry coefficient. For systems containing only a trace of permeant, the value of m may be a constant.

The Langmuir isotherm is often valid for adsorption of a monolayer of molecules on a surface or in pores. In many cases, more than only one monolayer of molecules can adsorb, and models for multilayer adsorption have to be used. For many polymeric materials, a combination of Henry and Langmuir sorption was found to apply (from the so-called free-volume theory). Many synthetic food packaging materials show this type of behaviour.

For edible materials, the interaction is often more complex, due to the complexity of the materials, which contain a combination of hydrophobic and hydrophilic groups, and usually show a clear microstructure. One often sees that so-called multilayer adsorption theories apply to these systems. The best-known examples are the Brunauer-Emmet-Teller (BET) isotherm (Bunauer *et al.*, 1938) and the Guggenheim-Anderson-De Boer relation (Weiser, 1985).

The parameters in the various models are usually determined by fitting the model to experimental data: measurement of the amount of moisture adsorbed by the matrix, when it is exposed to a certain activity of the permeant.

To know the concentrations valid for a barrier system, one has to consider three sorption isotherms: the one for the phase rich in permeant, the one for the phase poor in permeant, and the one for the barrier phase (see Table 5.2).

Although the previously mentioned models have successfully been applied to various foods, one needs to be careful when interpreting them. One pitfall is that the isotherm measured in a drying system, may be different from a re-hydrating system due to removal of water from pores that subsequently may collapse. This so-called hysteresis effect may even occur during the determination of the isotherm if the temperature is not very accurately controlled. Therefore, always try to find various sources for sorption isotherms and compare and be aware of this effect.

5.3. Kinetics: diffusion

It has already been mentioned that mass transfer can only take place if a driving force exists. This driving force results from a difference in thermodynamic potential or activity: the larger the driving force, the larger the transfer rate. During transfer,

Table 5.2. Models for adsorption isotherms[*].

Model	Equation	Parameters	Typical validity
Henry	$x_i^B = ma_i^A$	m	Simple liquids; dilute systems
Langmuir	$x_i^B = x_{i0} \dfrac{Ka_i^A}{1 + Ka_i^A}$	K - interaction coefficient x_{i0} - (maximum amount of sorption)	Monolayer adsorption; surfaces and porous systems; some glassy polymers
Dual Sorption	$x_i^B = ma_i^A + x_{i0} \dfrac{Ka_i^A}{1 + Ka_i^A}$	m – proportionality constant K – interaction coefficient x_{i0} – monolayer sorption content	Polymeric materials and solutions (combination of Henry and Langmuir sorption)
BET	$x_i^B = x_{i0} \dfrac{Ka_i^A}{\left(1 - a_i^A\right)\left(1 - (1-K)a_i^A\right)}$	K - interaction coefficient x_{i0} - (monolayer sorption content)	Multilayer adsorption; many food matrices
GAB	$x_i^B = x_{i0} \dfrac{KCa_i^A}{\left(1 - Ka_i^A\right)\left(1 - Ka_i^A + KCa_i^A\right)}$	K - multilayer coefficient C - Guggenheim constant x_{i0} - (monolayer sorption content)	Multilayer sorption; food matrices

[*]The models that relate the composition of component i and its activity a_i can also be expressed in terms of weight fractions (w_i) or volume fractions. The form of the equations does not change, although the numerical values of the parameters will change.

molecules experience a force balance. On the one hand, they experience the driving force pulling towards the region with lower chemical potential or activity. On the other hand, they are slowed down by friction with the environment. This friction depends on the difference in velocity between the molecule and its environment, and on the mole fraction of the other molecules in the environment.

The total force balance for the permeant (component 1) in a binary mixture of permeant and its environment (component 2) reads:

Friction force F_f = - Driving force F_z

$$\rightarrow \; \zeta_{12} x_2 u_1 = -\frac{d\mu_1}{dz} \tag{5.9}$$

in which u_1 is the velocity difference of the permeant with its environment and ζ_{12} is the friction coefficient between permeant and its environment.

Combination of Equation (5.9) with Equation (5.1) and rearrangement will yield the relative velocity of the permeant:

$$u_1 = -\left(\frac{RT}{\zeta_{12}}\right)\frac{1}{x_2}\frac{d\ln a_1}{dz} \quad (5.10)$$

The term (RT/ζ_{12}) is often called the (Maxwell-Stefan) diffusion coefficient Đ (m²/s). In a binary system $x_2=1-x_1$ and the general Maxwell-Stefan diffusion equation becomes:

$$u_1 = -\frac{Đ}{1-x_1}\frac{d\ln a_1}{dz} \quad (5.11)$$

This relation is expressed in terms of a relative *velocity*. The relation can also be expressed as *mass flux*. The mass flux is equal to $M_1 = \rho \cdot w_1 \cdot u_1$, where ρ is the mass density (kg/m³), w_1 is the weight fraction and M_1 is the amount of material passing through one m² of surface, per second (kg/m²·s).

$$M_1 = -\rho w_1 \frac{Đ}{1-x_1}\frac{d\ln a_1}{dz} \quad (5.12)$$

When the system is dilute in permeant (as is often the case in the barrier phase itself), the term

$1-x_1$ will be almost equal to 1 and the equation becomes:

$$M_1 = -\rho w_1 Đ\frac{d\ln a_1}{dz} \quad (5.13)$$

5.3.1. Incorporation of sorption in the transport equation

We cannot use this flux equation directly, because we mostly do not know the values of the activity coefficients in the barrier film. These are dependent on the sorption isotherm; we will look at three equations that may be used under different conditions: the Henry isotherm, the Langmuir isotherm, and the GAB isotherm. We will assume that component 1 is water.

Henry sorption

In the case of Henry sorption, the relation between the weight fraction and the activity of water is a linear relation: $w_1 = ma_1$. This can be entered in the flux equation, resulting in:

$$M_1 = -\rho ma_1 Đ\frac{d\ln a_1}{dz} = -\rho mĐ\frac{da_1}{dz} \quad (5.14)$$

If the diffusivity Đ and the Henry coefficient m are both constant, the difference in the water activity da_1 over the film at steady-state conditions is constant with the thickness of the layer (L). The equation can be represented by:

$$M_1 = -(\rho mĐ)\frac{\Delta a_1}{L} \quad (5.15)$$

The parameter $\rho\, m\text{Đ}$ (or $m\text{Đ}$, depending on the definition used for M_1) is called the *transfer coefficient* or *permeability coefficient*. It has unit m^2/s or $kg/m\cdot s$.

Permeability and permeance

Films are usually tested by exposing them at both sides with a gas (e.g. air), containing water vapour. At one side the gas is almost saturated with water vapour and at the other side the gas contains only a little water vapour. The water activity that is exposed (by the gas) to the film is equal to:

$$a_1 = \frac{p_1}{p_1^{sat}} \tag{5.16}$$

in which p_1 stands for the partial water vapour pressure, and p_1^{sat} the water saturation pressure. Entering this in the equation for the mass flux of the water gives:

$$M_1 = -\left(\frac{\rho\, m\text{Đ}}{p_1^{sat}}\right)\frac{\Delta p_1}{L} = -\frac{P}{L}\Delta p_1 \tag{5.17}$$

The symbol P stands for *permeability*, with unit $kg/m\cdot s\cdot Pa$. It plays an important role in the field of barrier technology. It is often assumed in literature that P is a material constant. It is however important to realise that the permeability is *only* a constant in the case of Henry sorption, and as long as the diffusivity is a constant. In all other cases, the permeability is not a constant, but can vary strongly, depending on the circumstances., also as a function of the position in a barrier.

In some cases, the thickness of a film is difficult to determine, and one simply uses the value of P/L, called the *permeance*. This value thus depends on the thickness of the film. Obviously, this is an average value, keep in mind that especially small defects will influence this value largely (see also Section 5.6).

It's a sure sign that Henry adsorption and the requirement of a constant diffusion coefficient are not valid, when you see reported values of the permeability that are dependent on the gradient of water vapour pressure that is applied over the film. This does not mean that you cannot use the reported values, but you should be careful to use the data under the reported conditions, or to take a wide safety margin.

We will now show how the relations change, if the film material shows sorption behaviour that is different from Henry behaviour.

Langmuir and GAB sorption

The equation for Langmuir sorption $w_1 = w_{10}Ka_1/(1+Ka_1)$ can be entered in the mass flux equation (Equation 5.13):

$$M_1 = -\rho\frac{w_{10}Ka_1}{1+Ka_1}\text{Đ}\frac{d\ln a_1}{dz} = -\rho\frac{w_{10}K}{1+Ka_1}\text{Đ}\frac{da_1}{dz} \tag{5.18}$$

At small values of a_1, the type of relation is the same as for Henry. However, at higher values of a_1, the relation becomes different and this will have consequences for the permeability:

$$M_1 = -\rho \frac{w_{10}K}{(1 + Ka_1)p_1^{sat}} Ð \frac{dp_1}{dz} \approx P \frac{dp_1}{dz} \qquad (5.19)$$

It is obvious that at higher values of a_1 the permeability P is no longer a constant, but instead decreases with increasing water activity a_1. Note that we cannot easily integrate over the thickness of the film, as P is dependent on the local water activity.

For GAB sorption a similar relation can be found:

$$M_1 = -\left(\rho \frac{w_{10}KC}{(1 - Ka_1)(1 - Ka_1 + KCa_1)p^{sat}} \right) \frac{dp_1}{dz} \qquad (5.20)$$

With GAB adsorption, the permeability P may in fact *increase* with increasing water activity.

What is the impact of this on possible fluxes through the films? Let us investigate the concentration gradients that may occur in a film in more detail. We assume that we have a film that is exposed at one side to a water activity of 0.8, and at the other side to a water activity of 0. We assume further that we have three different materials with different sorption behaviour. One shows Henry adsorption, one shows Langmuir adsorption and one shows GAB adsorption. The isotherms are equivalent in the sense that at water activities 0 and 0.8, all three show the same amount of water sorption: they seem to behave similarly (Figure 5.1).

With this figure for the mass of moisture adsorbed, the water activity at each position in the film can be calculated (Figure 5.2a) and via the adsorption isotherm, the amount of water adsorbed in the film at each position (Figure 5.2b) in case of Henry adsorption, Langmuir adsorption and GAB adsorption. It is obvious that the type of adsorption isotherm results in completely different concentrations and water activity profiles in the film. The water fluxes in the barrier material will be very different for each type of barrier material, since the water activity gradient (and concentration gradient) differs for each material.

The driving force for mass transfer through the barrier strongly depends on the type of adsorption behaviour. The water flux in the barrier material which shows Langmuir behaviour will be 200% higher than the water flux in the barrier that shows Henry behaviour. The water flux in the barrier with the GAB isotherm shows a water flux that is only 26% of the Henry behaviour, even though the overall value of the sorption at both sides is taken to be equal.

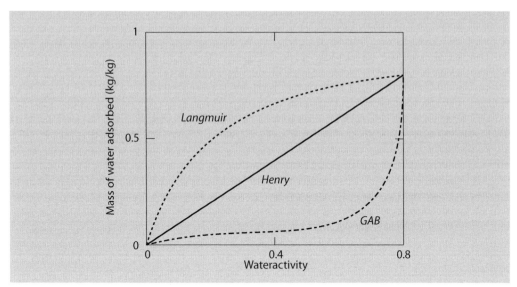

Figure 5.1. Sorption isotherms of water at different water activity for three different materials. Henry adsorption (for m = 1); Langmuir adsorption (for K = 5 and w_{w0} = 1), and GAB adsorption (for w_{w0} = 0.03205, K = 1.2 and C = 26).

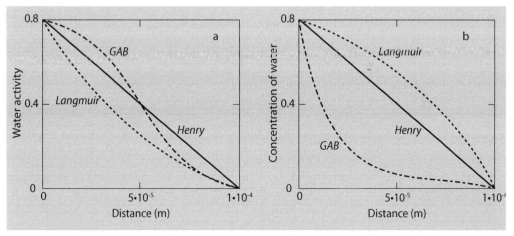

Figure 5.2. (a) Water activity profiles in the barrier film. (b) Concentration profiles in the barrier film in three different barrier materials. Henry adsorption (for m = 1); Langmuir adsorption (for K = 5 and w_{w0} = 1), and GAB adsorption (for w_{w0} = 0.03205, K = 1.2 and C = 26).

It is therefore clear that the type of isotherm should be taken into account, and that calculations based on a constant overall permeability coefficient lead to erroneous results in estimated fluxes.

Use of adsorption data from literature

When using data from literature, always check the circumstances under which the measurements were taken. The literature data can be used if one applies the film under the conditions that were used during the measurements reported (water activities at both sides). If the literature data show that the permeability is independent of the water activity gradient applied, you can also use the literature data, as this is an indication that Henry sorption applies, and the diffusivity is approximately constant. If this is not the case, but you only need a rough estimate, you can use the overall permeability, but should check how reliable your result is. Remember from the graphs above that misinterpreting the type of sorption isotherm may lead to an error larger than an order of magnitude!

5.3.2. Influence of the thickness of the film

From the above discussion it will be clear that the thinner the film, the higher the mass flux through the film. If the permeability P is constant, one would expect that halving the thickness of the film would result in doubling of the mass flux. However, in many cases, the total flux through the membrane is *not* linearly decreasing with the thickness of the film. The flux may vary with a factor Δz^n in which n may vary from 0.8 to 2.0 (Figure 5.3). For a film of 0.01 mm thickness a water vapour permeance that ranges from 63 to 158 g/m^2 h kPa can be calculated, based on literature data for the same type of film with a thickness of 0.1 mm and a permeance of 10 g/m^2 h kPa. It is not clear what causes this non-linearity, but it could be due to heterogeneities in the barrier (see Section 5.5). This example just shows that practice does not always follow theory completely.

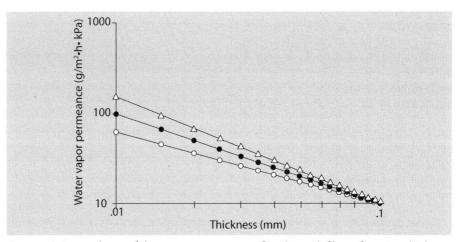

Figure 5.3. Dependence of the water-vapour mass flux through films of varying thickness Δ n=0.8; ● n = 1, o n=1.2 (Chen, 1995).

5.3.3. Values for the Maxwell-Stefan diffusivity of film materials

Diffusivities are extensively tabulated in handbooks. One of the reasons for this is that it is generally not yet possible to predict values for diffusion coefficients reliably. It is however possible to give some rough guidelines, for different classes of materials.

Gases	$\sim 10^{-5}$ m^2/s
Liquids	$\sim 10^{-9}$ m^2/s
Rubbery polymeric materials	$10^{-14} - 10^{-10}$ m^2/s
Glassy polymeric materials	$10^{-16} - 10^{-12}$ m^2/s
Crystalline polymers	$10^{-18} - 10^{-14}$ m^2/s
Lipid materials	$\sim 10^{-12}$ m^2/s
(chocolate	$\sim 10^{-13}$ m^2/s)
Protein films	$\sim 10^{-13}$ m^2/s
Polysaccharide films	$\sim 10^{-13}$ m^2/s
LDPE (comparison)	$\sim 10^{-12}$ m^2/s

The diffusivity in polymeric materials is generally strongly dependent on the amount of permeant in the matrix Figure 5.4). In the case of water permeation, the diffusion coefficient of water in the dry material is extremely low (between 10^{-18} and 10^{-16} m^2/s). It then rises more or less exponentially with the amount of water in the

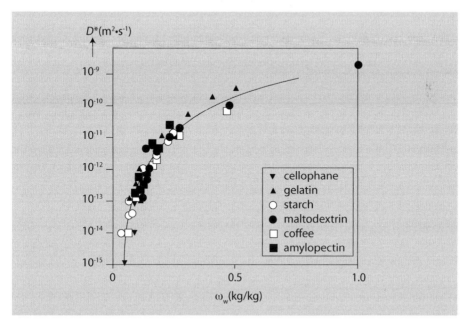

Figure 5.4. Effective diffusion coefficient of water as function of its weight fraction, found to be valid for a range of hydrophilic materials (cellophane, gelatine, starch, maltodextrin, coffee, amylopectin) (Bruin, 1980).

film material; at 1 wt% of water in the film matrix, the value of the diffusivity can already be 10^{-13} m^2/s; at 5 wt% this can have risen to around 10^{-10} m^2/s.

The diffusivity of water in non-polymeric media (such as lipids) is usually less dependent on the amount of water present, especially when using the activity-based transport equation, as we did. However, if a material is glassy, the diffusivity will be extremely low, too low to be measured.

Please note that if a material is glassy (and stays glassy), the diffusivity will be extremely low, too low to be measured. If a small amount of water is absorbed in the material, the glass transition may be crossed, and the diffusivity suddenly rises to normal values. One can expect this behaviour with carbohydrates in particular. Even though it seems that this glassy behaviour renders these materials suitable as barrier materials, the possible crossing of the glass transition would make the barrier worthless. They can therefore only be used when the water activities remain low at both sides of the film. Further, glassy materials usually are brittle, which makes them fragile and difficult to handle.

5.4. Barrier technology: practical aspects and guidelines

As discussed in the previous sections, the basis of the design of a barrier lies in thermodynamics (chemical potential differences, sorption isotherms and driving forces) and in dynamics (mobility and diffusivity). These two domains are the basis on which the design of barriers in food is founded. However, in the design of a suitable barrier in structured food products we often encounter practical situations that need to be taken into account.

Here we will provide practical guidelines on how appropriate coating materials and a suitable application system or coating process can be selected for a specific purpose given a combination of constraints. It includes a discussion on properties that can be expected with various barrier materials, e.g. protein or polysaccharide, lipophilic or amphiphilic barriers. The application of heterogeneous coatings will be briefly discussed, as will the relevance of achieving defect-free coatings.

5.5. Heterogeneous films

Many practical coatings are in fact not homogeneous. Many polymeric coatings are semi-crystalline; some lipid-based films may be highly crystalline, and many protein-based films have a complex structure, due to the presence of more than one protein and often other components. The structure of the coating, however, strongly influences the permeability of the film.

It is difficult to capture all the effects of heterogeneous films. However some basic rules can be given. There must first be information on the structure of the film. There are a number of simplified structures that can be taken as a basis (Figure 5.5). We will assume the presence of two phases, 1 and 2.

- The series model assumes that the two phases are situated as layers normal to the direction of diffusion (i.e. parallel to the surface). Here, the overall permeability depends on the volume fraction φ of the phases and the permeability P of the individual phases:

$$\frac{1}{P} = \frac{1-\phi_2}{P_1} + \frac{\phi_2}{P_2} \qquad (5.21)$$

This model is applicable, for example, for films that contain plate-like crystals that are oriented in the direction mentioned. One can see that the crystalline fraction in this case has a very strong influence on the permeability.

- The parallel model assumes that the two phases are placed in layers parallel to the direction of permeant diffusion. The permeability is then given by:

$$P = (1-\phi_2)P_1 + \phi_2 P_2 \qquad (5.22)$$

In general, this situation is not desirable in a barrier; the more permeable component will determine the overall permeability of the barrier, although the effective surface area through which permeation can take place may be reduced considerably.

- The Maxwell model assumes that a dispersed phase is present as a random packing of uniform spheres. The diffusivity is then given by:

$$P = P_1 \frac{2P_1 + P_2 - 2\phi_2(P_1 - P_2)}{2P_1 + P_2 + \phi_2(P_1 - P_2)} \qquad (5.23)$$

Let us consider a lipid film that has a crystalline fraction of 80%. The crystals are more or less randomly distributed in the film. If we assume that the crystals can

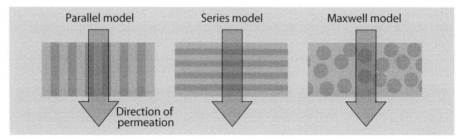

Figure 5.5. Three models for in-homogenous layers: layers in series, layers in parallel, and a dispersion of equally-sized spheres of one phase in the other.

be interpreted as spheres, we can use the last model, which is a good assumption for many materials. The diffusivity inside a crystal is extremely small, and we may take it to be zero, compared to the diffusivity of the non-crystalline matrix around it. The resulting diffusivity that we estimate with the Maxwell model, is only 14.3 % of the diffusivity of the non-crystalline phase.

Apparently, a large crystalline fraction seems to be a good property of a barrier film, as it significantly reduces the permeability by effectively increasing the diffusion distance. However, a high crystalline fraction may also influence the mechanical behaviour: if the crystalline fraction is very high, the film may become brittle, and may show defects. In addition, if you have a crystalline lipid film, and you need to subject the total product to a temperature treatment, the film may melt and loose its coherence (Figure 5.6).

5.6. Prevention of defects in barrier films

5.6.1. Defects in barrier films cause severe problems

Let us consider a barrier film made of paraffin that has a typical permeability of around $0.2 \cdot 10^{-15}$ kg/m·s·Pa. Let us consider using a film with thickness 100 μm, and let us assume that at one side of the film we have a completely dry compartment (0 % relative humidity), and at the other side we have a compartment with water activity 1 (100% relative humidity).

At 25 °C and atmospheric pressure we can calculate with the Antoine equation, that $a_w = 1$ corresponds with a partial water vapour pressure of 3.158 kPa, or, air that is completely saturated with water vapour consists of $3.158 \cdot 10^3 / 10^5 = 3.2$ % of water

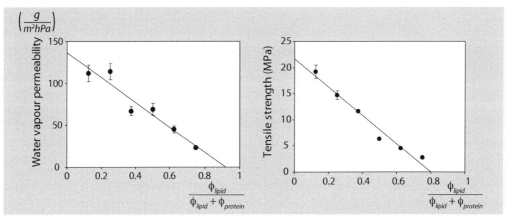

Figure 5.6. The effect of using an emulsion: one may reduce the water vapour permeability of a protein coating by adding a lipid, but at the cost of mechanical strength (Chen, 1994).

molecules. This leads to a flux of water through the film of: $M = - (P/L)\cdot(p_0-p_1) = -(0.2\cdot10^{-15}/100\cdot10^{-6})\cdot(0-3158) = 6.3\cdot10^{-9}$ kg/m²·s $(= 0.55$ g/day·m²).

But what happens if the film contains a defect? The flow of water vapour through the defect is given by:

$$M = -\rho_a \frac{Đ_a}{L}\left(y_{w0} - y_{w1}\right) = -\frac{P_a}{L}\left(p_{w0} - p_{w1}\right) \tag{5.24}$$

in which ρ_a is the density of air (1.185 kg/m³), $Đ_a$ is the diffusivity of water vapour in air at 25 °C (2.56·10⁻⁵ m²/s), y_{wi} is the molar fraction of water in the air at both sides of the film, and L is the thickness of the film.

The 'permeability' P_a of a defect containing air is equal to $(\rho_a \, Đ_a) / p_t$ in which p_t is the total pressure (10⁵ Pa). Its value is calculated to be 3.0·10⁻¹⁰ kg/m·s·Pa, which is more than one million times higher than the permeability of the film itself!

The resulting total permeability can now be calculated by assuming that the defect is a cylindrical pore straight through the film. In that case we may use the plates-in-parallel model: the permeation through the film and through the defect take place in parallel $P = (1-\phi_a) \, P_f + \phi_a P_a$.

With this relation we find that even when we only have a defect as small as 6 millionths of the total surface ($\phi_a = 6 \cdot 10^{-6}$), the effective permeability has already gone up a factor 10 (Figure 5.7)! It is obviously of the utmost importance to avoid the formation of any defects. In an industrial situation, these films have to be added to hundreds of products per minute, often under non-ideal circumstances in factories around the world, and have to retain their integrity during transport; clearly, this is a major challenge.

5.6.2. Some solutions to avoid defects in coatings

Applying thick coatings

There are a number of possible approaches to avoid defects in coatings. The most obvious one is of course to use a thick film. It is always possible to make a film free of defects by applying a thick layer. The disadvantages are obvious: one needs a lot of coating material, which may be expensive. It may affect the taste and visual appearance as well as the tactile and olfactory characteristics of the product. Additionally, a thick coating may need a complex application process: the conditions need to be such that the coating material is not drained from the product in the time it needs to solidify. Since the speed of drainage of a film from a product depends on the 4th power of the thickness, a film that is twice as thick leads to a drainage time that is 8 times shorter (even though the time needed to solidify will be more than twice as long). Therefore, there are practical limits to the thickness of the film, even when consumer perception or product or process costs are not an issue.

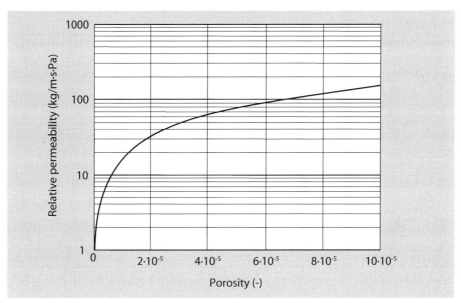

Figure 5.7. Effect of small defects on the total permeability of a paraffin film. At a porosity of 0.01%, the effective permeability has gone up more than a factor 150.

An example is the coating of many products with chocolate. Since the coating procedure usually takes place from the melt, the solidification process by crystallisation does not take too long. In addition, the viscosity of a chocolate melt can be relatively high, increasing the drainage time of the chocolate film. However, it is clear that the presence of the chocolate layer will be dominant in the perception of the consumer; the design of the coating is therefore central in the design of the product.

Using emulsions

The use of emulsions may be a good way to reduce the formation of cracks. After the drying of a lipid emulsion stabilised by a protein, a relatively flexible continuous phase encloses the lipid globules, thereby forming a robust film. Although in this way a more or less defect-free film is obtained, a substantial volume fraction of the continuous phase is always needed, in order to maintain the matrix flexibility and tensile strength. Unfortunately, a protein- (or polysaccharide) based continuous phase features a much higher permeability for water. Use of the Maxwell model shows that this inevitably leads to relatively high water vapour permeability (at relatively low ϕ_2).

Applying multiple layers of the same material

An obvious solution to the problem of defects in films, is applying multiple layers on top of each other (Figure 5.8). Even though each layer may have some defects,

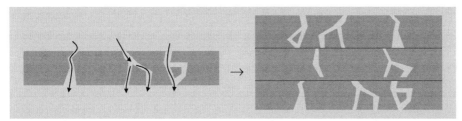

Figure 5.8. Application of multiple films may lead to closure of various defects, giving a defect-free system.

it is unlikely that defects will occur at the same place, and thus, the strong short-circuiting effects that we discussed previously will be greatly reduced. To overcome these problems, one typically has to apply the layers several times (4-10 times).

This solution has the advantage that it is not limited by the drainage speed of the liquid coating material: each layer can be thin, and a solidification step (e.g. by evaporation) in between each coating step may give the total system sufficient strength during preparation. Of course, the total barrier may still be rather thick, which has implications for consumer perception similar to the application of one thick coating. For this route in particular, the costs may quickly become unrealistic: if 5 layers need to be applied before an adequately defect-free barrier is obtained, one needs 5 times the coating and solidification process steps and the related additional processing time.

Application of a combination of a low and high permeable film

An interesting route is the use of a bi-layer system. A first layer is applied of a material that has a very low permeability to the undesired permeant (such as water). This may, for instance, be a lipid based film, featuring very low permeabilities, but also a low mechanical strength. Thus, it is almost inevitable that this film will have defects. Then, a second coating is applied over this layer. This is made of a relatively permeable material, but has optimal mechanical properties, such as high elasticity and high tensile strength. This might be a protein-based coating. The thickness of this film may be low; it is essential, however, that the coating material wets the previous layer. If it does, this second layer will penetrate the defects through capillary action, and fill it with the second coating material.

This leads to a 'plugging' of the defects (Figure 5.9). Let us use the earlier example (a paraffin wax based primary coating), but now apply a second coating with a protein-based film with relatively high permeability of $36 \cdot 10^{-13}$ kg/m·s·Pa. This material penetrates the defects. At a defect volume fraction of 0.1%, the effective permeability is now only 20 times higher than the intrinsic value of paraffin wax, instead of 1500 times, when no secondary coating is applied.

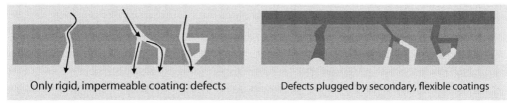

Only rigid, impermeable coating: defects   Defects plugged by secondary, flexible coatings

Figure 5.9. The idea of plugging defects with a flexible secondary coating, which has higher permeability.

In Figure 5.10 one can clearly see that it is not necessary for the secondary coating to have very good barrier properties. A much lower permeability in this layer results in a small reduction of the overall permeability of the barrier. With this technique, one can therefore overcome the traditional dilemma that most highly impermeable materials have poor mechanical and rheological properties, and *vice versa*.

There are also disadvantages: the primary layer should still be applied with reasonable effectiveness (the porosity should not be too high) and more importantly, the second layer should have excellent binding properties to the material of the primary layer. The wetting behaviour therefore becomes crucial. And two coating procedures are still needed, which incurs relatively high costs.

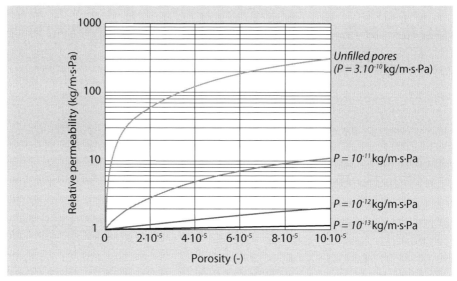

Figure 5.10. The influence of various types of secondary coatings. One can see that the permeability of the secondary coating is not very important. A value of 10^{-11} kg/m·s·Pa is unrealistically high; most highly permeable materials are below 10^{-12} kg/m·s·Pa.

     Food product design

Curing

A third possible solution to prevent defects in a coating is to cure a coating, once applied. If the material is brought into a condition or temperature, where it can deform slowly (creep), the capillary pressure may induce closure of the defects. Unfortunately, this only works for very small defects, whereas the larger defects cause most of the permeability problems (Figure 5.11).

If there are no serious problems with defects, curing can have an additional effect. Due to Ostwald ripening and re-crystallisation, the crystals will become bigger, and crystallisation will become more complete. In some cases, the transition to a different form of crystal may have a beneficial effect. Stearyl alcohol, for example, was found to show a decrease in permeability when cured for 0, 14 and 35 days at 49 °C, due to the formation of larger plate-like crystals, parallel to the surface (see section on heterogeneous films).

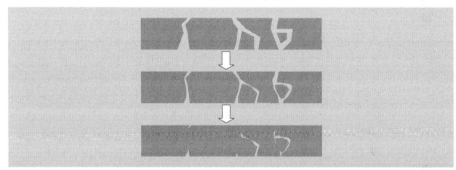

Figure 5.11. Through curing, pores will be slowly filled by creep flow, due to capillary effects.

5.7. Guidelines for application of coatings

A short overview will be given; a more comprehensive treatment can be found in the literature (Krochta, 1994).

5.7.1. Ingredients

Biopolymers

The most important biopolymers include proteins (e.g. gelatine, wheat gluten, soya protein, whey proteins, casein and zein), cellulose derivates, plant gums (alginates, pectins, gum Arabic), starches (possibly chemically modified) or other polysaccharides.

Most biopolymers are hydrophilic in nature (i.e. they have strong interaction with water, and often even dissolve in it). This implies that the solubility of water in those materials is very high. It will be clear that this results in a rather high moisture permeability. Therefore, one would preferentially not use them for the design of a moisture barrier. However, they do have good barrier properties regarding oxygen, carbon dioxide and lipid components. Furthermore, their mechanical properties are very good, especially when combined with a plasticizer. They combine good film elasticity (tensile strength) with a good elongation at break. In some applications, the water solubility of polysaccharides can be made use of, e.g. in case the film should disappear during the use of the product.

An important property of biopolymers is the charge of these molecules. By changing the pH it is possible to alter both the permeability and the mechanical properties of the material. In general, when the biopolymer is near the iso-electric point (charge neutral), its behaviour will be more hydrophobic, while its mechanical strength will be less.

Lipids

Since water is poorly soluble in lipids, they are very suitable as a barrier against moisture migration; however, their mechanical strength is poor compared to those of a biopolymer-based coating. Waxes and paraffins are often used on fruit to decrease respiration and moisture loss. Acetylated monoglycerides are often added to slightly improve the poor mechanical properties and wetting behaviour.

Many lipids are crystalline. The extremely low solubility and diffusivity in crystals lead to a low overall permeability. The form and orientation of crystals can be important to the permeability behaviour: planar crystals oriented parallel to the surface act as individual layers in series. This leads to a much lower overall permeability. However, the higher the crystalline fraction, the poorer the mechanical properties.

Emulsions and composites

Lipids have very good barrier properties, but may be difficult to apply, and have poor mechanical properties when applied in isolation. That is why they have been incorporated in emulsions, where they are dispersed with bio-surfactants (e.g. proteins). More components are usually added to change the rheological behaviour (see above for its relevance to coating).

These emulsions can then be sprayed on the product. Again, the solvent, usually water or a water/ethanol mixture, has to be evaporated, and once again, multiple coating steps may be necessary to obtain a film that is sufficiently defect-free.

The drying conditions are highly important for both the homogeneous and emulsion casting techniques. Drying at high temperatures may lead to creaming and coalescence of the fat globules, forming a homogeneous layer on top of the coating. This will in turn reduce the drying rate substantially, as it prevents the transport of water to the vapour phase. The coating may over-heat locally and defects may be formed, leading ultimately to a non-homogeneous, defective coating. Adding components such as lecithin, to reduce the tendency to cream, can reduce these effects.

Additives

Additives are used almost without exception. When using a biopolymer, a plasticizer is often needed. Without this plasticizer, the film would be brittle, fragile and glassy, which makes it difficult to prepare a defect-free film, but which also makes the product difficult to handle. A plasticizer is a component that is present in smaller concentrations (5-15 wt%), and which changes the mechanical properties. Typical plasticizers are water (an excellent plasticizer, but not practical in a moisture barrier), glycerol and polyethylene glycols of various molecular weights. Addition of these components changes the properties from glassy and brittle to pliable and elastic, often called rubbery. Too much plasticizer leads to loss of mechanical strength and ultimately loss of coherence of the layer.

Plasticizers for biopolymers are hydrophilic in nature, which means that their addition to the film will lead to an increase in moisture permeability, thereby compromising on the quality of the barrier function. Furthermore, it should be noted that plasticizers are usually relatively small molecules that may have good interaction with the product substrate. Thus, the plasticizer may slowly migrate from the film into the product, leading to a slow loss of elasticity of the film, and subsequent failure of the film, and possibly deterioration of the product itself.

Other additives are also applied, such as antimicrobial agents, vitamins, antioxidants, flavours and colorants. The use of these additives adds other functionalities to the film.

5.7.2. Formation mechanisms

Melt-based formation mechanism

For those materials that can form a melt at attainable temperatures, melt-based casting is a very good option. Here, the product is dipped in a bath of the melt. An obvious example of such a process is the preparation of chocolate coatings; please note that it is not trivial to coat a product all around, and often more steps are needed to e.g. coat top/sides, and bottom of a product.

Two mechanisms may lead to the formation of a coating. First, if the product itself is below the crystallisation or solidification temperature, the melt will solidify against the product. If this is caused by crystallisation, e.g. in chocolate coating, this will lead to a solidified layer around the product (Figure 5.12). The rate of increase of thickness of

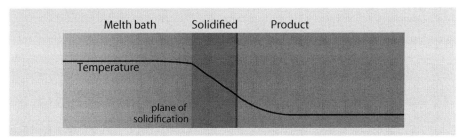

Figure 5.12. The melt solidifies against the product, due to heat transferred to the cold product. The longer the product is immersed, the thicker the coating will get.

the layer depends on a number of things, which we can find from a simple heat balance. The thickness of the layer increases with the square root of time: if the product stays twice as long in the bath, the layer will become approximately 1.4 times as thick.

A second possibility is that the melt does not solidify during dipping; a layer of liquid material sticks to the surface of the product. In this case, the time of immersion is not important. The product should be transferred as quickly as possible to a cooler (e.g. a blast cooler), where the film is solidified. Again, the time one needs to keep the product in the cooler is dependent on the square of the thickness of the coating. A coating that is twice as thick, needs a cooling time that is 4 times as long. When the coating melt has approximately Newtonian behaviour, the liquid will drain from the product as soon as it is taken from the bath (Figure 5.13). The amount of material remaining on the product is then dependent on the inverse of the square root of time.

Rotating the product, thereby preventing the material dripping off before it has solidified, can reduce this effect. The coating material may show non-Newtonian shear-thinning behaviour and possibly even have a yield stress. The liquid film on the product will then be much more stable. One can achieve these effects by adding components, such as polymers, to the solution.

The properties of the film can be altered by tempering (annealing at higher temperatures). This can lead to re-crystallisation, giving other crystal shapes or a different orientation in the film. It can also lead to a reduction in the number of defects and a change in mechanical properties, and in different consumer perceptions.

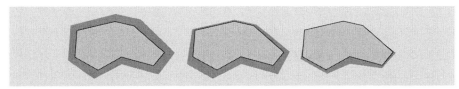

Figure 5.13. If the coating solution does not have a sufficiently high viscosity or has another barrier against flow, the coating may become inhomogeneous due to film drainage.

Solvent evaporation

Solvent-based casting (wet casting) is another film formation mechanism. It uses a solution of a coating material instead of a melt. One can either spray the solution on the product, which is the most commonly used method, or dip the product in the solution.

By influencing the rheology of the solution, the thickness of the coating can be regulated; by adding surface-active components, one can influence the spreadability of the coating. By taking a mixture of water and ethanol as solvent, the evaporation process can be speeded up; the first stages of evaporation are particularly important, as the solution is still diluted then, and may migrate, by drainage or by capillary action of the product (Figure 5.14). The disadvantage of this method is that the solvent has to evaporate; this means that an additional drying step is necessary, which implies that you need to consider the related processing time, equipment, and energy costs. If the wetting of the solution to the product is not good, several spraying steps may be necessary, with all its consequences.

It is obvious that this method is not applicable where the product substrate is sensitive to moisture. When the substrate is a cracker, spraying of a water-based solution will lead to a loss in crispness, and thus not fulfil its goal. One of the solutions may be to consider a non-aqueous (food grade) system, but this may lead to residues which are perhaps unwanted. In this case, other systems may be more appropriate.

Coacervation

Coacervation has been applied in the pharmaceutical area, but not to any great degree in the food industry. It involves the creation of an insoluble, usually polymeric substance (the so-called coacervate), which precipitates onto an available surface (such as the product surface). This layer can be formed by reducing the solubility of one polymer, e.g. through adjusting the pH or the temperature or adding a non-solvent or salts. Another method is the application of two polymers with opposite charges,

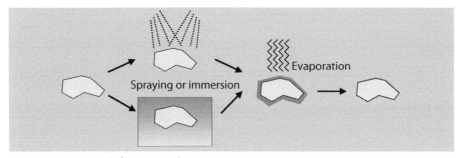

Figure 5.14. Methods for using solvent evaporation, via dipping or spraying.

which together form a precipitate. This can be done with aqueous or non-aqueous solutions (Poncelet, 2006).

5.7.3. Application methods

Mechanical method

Everyone is familiar with the use of a knife to put butter on a slice of bread before adding some jelly. This habit is born from the wish to prevent the jelly migrating into the bread and changing the physical properties as well as the perception of the product. This mechanical technique is used for a limited number of applications.

One of the mechanical methods used to form an isolated film (i.e. not supported by a product substrate) used on an industrial scale, is the creation of a well-defined layer of a solution or melt on a smooth surface, with a casting knife that may have a solution reservoir. The knife forms a layer of solution on the surface of the product. By subsequent drying of cooling (sometimes combined with cross linking), a film is formed (Figure 5.15).

Other ways to apply a coating are with the help of brushes (just like painting), or by letting the product pass a falling film, which then spreads around the product. For flat surfaces, rollers can be used.

Dip casting

This method is often applied for melts (see above), but it can also be applied to solutions. The product is immersed in a bath, during which a film is formed (Figure 5.16). Solidification can take place during immersion (melt-based) or afterwards in an evaporation or cooling step.

It is important that the substrate and the film material have good surface compatibility. If the surface is not wetted, the film will not be formed properly; or, if the surface has a very strong interaction with the solution (e.g. a cracker with an aqueous solution), the solution may penetrate too deep in the product, leading to product deterioration and the formation of bad films.

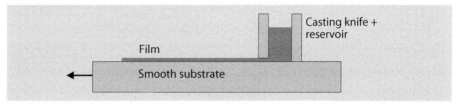

Figure 5.15. Knife-based casting, with reservoir built inside the knife. The design of the knife determines the thickness of the casting layer.

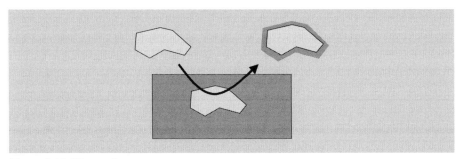

Figure 5.16. Dip casting.

Spraying

Just like spray painting, spraying can lead to well-defined, thin films, while the product can be sprayed from all sides, or specifically from one side (e.g. a pizza exposed to a high-moisture filling). Usually a solution is used, although there are examples of melt-based spraying. It is not difficult to use this method for the formation of a number of layers, for example, to achieve a good defect-free barrier, or to combine low-moisture permeance with low oxygen or low lipid permeance.

5.8. Mechanical behaviour

A coating that encloses a (part of a) product becomes an integral part of that product. A product is generally subject to changes in size and shape, due to temperature changes, due to possible losses in water content, and also by mechanical handling, if the product is a 'soft solid'.

A well-known example is the difficulty of coating an ice-cream product with chocolate: during processing (cooling and freezing), handling and transportation, the product will be subject to changes in temperature. This will cause the ice cream (and coating) to change in size: when the temperature of the ice cream decreases, some of the (highly-concentrated) solution will crystallise further, leading to an increase in volume. The chocolate around the ice cream has the tendency to shrink with lower temperatures. Tensions build up in the coating and cracks may appear. Therefore, designing a chocolate coating around such a product that remains defect-free is no easy matter. The fact that there are crack-free chocolate-coated ice creams on the market shows that it can be achieved.

All products undergo volume changes when subjected to a variation in temperature. The coating should have approximately the same volumetric expansion coefficient: it should shrink or expand together with the product, if the coating is not flexible. This is especially important during transport (either to the shop or to your home) since temperature is in general less controlled under such conditions.

When a product is not rigid, but may change in shape, the coating needs to be flexible. In particular, coatings that are based on polymers (e.g. proteins) can be adjusted to meet the flexibility demand. Although pure proteins or polysaccharides can be rather brittle or even glassy (implying that the elongation at break is extremely small, not higher than 1 or 2 percent), their behaviour can be influenced by using so-called plasticizers: usually small molecules, which mix well with the polymers, and make them swell and somewhat pliable (Figure 5.17). The more plasticizer is applied, the softer the material becomes, until it becomes basically a liquid. The best known plasticizer for proteins or polysaccharides is water. However, water has the tendency to migrate to locations of low water activity. In addition, it has the tendency to evaporate. If this happens, the film will at the very least lose its flexibility and the coating may fail. Therefore, other components are often used, such as glycerol or polyethylene glycol oligomers (e.g. polyethylene glycol 400), which are not volatile, and have different interactions with many food materials. The larger the molecular weight of the plasticizer, the slower its migration to other places.

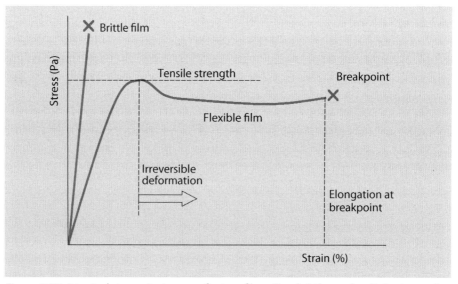

Figure 5.17. A typical stress-strain curve for two films. One brittle – only a little elongation is possible before the film breaks; the other more flexible – after some elastic deformation, the film may be permanently deformed when the strain is further increased. This permanent deformation may also influence the permeation properties, as it may induce the formation of defects. One of the ways to make a brittle material more flexible is to incorporate some plasticizer.

5.9. Wetting behaviour

5.9.1. Smooth surfaces

In the application of thin films, surface interactions always play an important role. The solution or melt applied should more or less form a film. This can only be achieved if wetting between the two materials (substrate and coating solution) occurs.

In general, a good film on a smooth surface is formed if the contact angle θ between the substrate and the coating is low. When the contact angle is zero, the spreading will be spontaneous. If it is sufficiently small, but at least smaller than 90°, it will be possible to form a smooth film: forced wetting of the surface takes place. If the contact angle is larger, the solution will have the tendency to form individual drops on the surface, impeding the formation of a smooth film (Figure 5.18).

The contact angle can be measured experimentally, but it can also be derived from the interfacial tensions γ between the surfaces: γ_{GS} surface(S)/air(G), γ_{LS} coating solution(L)/surface and γ_{LG} coating solution/air, with the help of Young's equation. In general, these interfacial tensions are hardly tabulated. Therefore, experimental measurement is the fastest way to a reliable estimation of the wetting behaviour. Besides, the effect of a surfactant, that will influence all γ's in Equation 5.25, and therewith the contact angle, can be assessed.

$$\gamma_{LG} \cos\theta = \gamma_{GS} - \gamma_{LS} \tag{5.25}$$

5.9.2. Rough and porous surfaces

Most product surfaces are not smooth. In fact, many products are highly porous. This greatly influences the film formation. A rough surface may help the casting solution to stick to the surface, as long as the contact angle is smaller than 90°, and the roughness is not too pronounced.

After application of the coating and drying, the surface will be coated. However, the ends of the corrugations may still stick out of the coating. Since the product is much more permeable to the permeant than the coating material, this will lead to defects in the coating, which will destroy any barrier functionality. Often, the coating itself

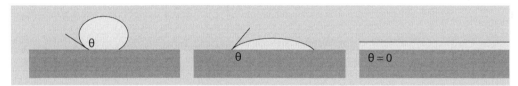

Figure 5.18. Non-wetting, wetting, and spreading.

is regarded as a good substrate for a second coating layer: the coating material will probably have a contact angle that is lower than 90°.[2]

If the initial contact angle is more than 90° (i.e. the surface is not wetted), the coating solution droplets may not penetrate between the surface corrugations, and will even be repelled from the surface (Figure 5.19). In this case, the formation of a film is almost impossible. However, by applying surfactants, it could be possible to reduce the contact angle sufficiently to allow wetting.

When the surface is porous (e.g. crackers and cookies), the coating solution may penetrate deep into the sub layer, 'sucking' too much solution away from the surface by capillary action, and therefore impeding the formation of a good film. When using a dispersion or emulsion, one can actually make use of this phenomenon: while the solvent is drained away into the pores, the emulsion or dispersion droplets that remain at the surface are able to form a dense layer covering the product. Of course this means that the concentration of moisture in the sub layer will increase somewhat, which may have negative effects on its quality, unless this effect is accounted for. Therefore, the precise properties of the product (sorption isotherm, plus for example the boundaries of the glass region) and its shape and size will determine whether this is a feasible option or not.

For rough and porous systems, spraying is the only acceptable coating procedure that allows control over the layer thickness. Often, the whole surface area cannot be covered in one single coating step, especially if the sub layer is porous. Several steps may be necessary before the openings of the pores are completely sealed. Of course this depends very much on the porosity and size of the pores.

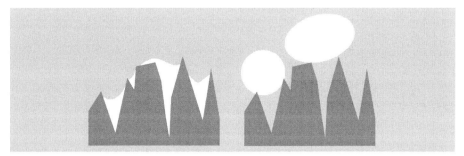

Figure 5.19. Surface roughness will help the formation of a coating as long as the contact angle is less than 90°. If it is greater than 90°, roughness will prevent the coating from sticking, and no adequate coating will be formed.

---

[2] This may not be true for proteins, which may develop a hydrophobic surface during drying. Also, the use of emulsions or dispersions with a hydrophobic material may lead to segregation, creaming and the formation of a hydrophobic surface.

5.10. Migration of other components

5.10.1. Unwanted migration

Coatings are often made of materials that are rather different from the components in the product to be coated. This may induce unwanted migration of components. For example, a film meant to prevent moisture migration may consist at least partly of lipophilic components. Lipids that may be present in the product and are in contact with the coating can migrate from the product to the coating and accumulate in the coating. This can lead to a change in properties and consumer perception of the product, and even failure of the coating, leading to further quality loss.

An example of the first effect is found in the case of a chocolate coating. Fat migration can lead to 'blooming' (fat redistribution) of the chocolate leading to a less attractive appearance. Another example is the migration of flavour components into the coating, thus depleting the product of these important constituents. Migration of colorants, natural or added, may lead to a distorted appearance and loss of barrier functionality of the product.

Especially in cases where the product contains a large amount of migrating species, this may lead to swelling of the coating, build-up of internal mechanical stresses and ultimately to the formation of cracks with all its consequences for the barrier function (see section 5.6). It is therefore important that the composition of the coating matches the product composition and morphology.

Although it is difficult to predict how components will migrate, in general components with the same degree of hydrophilicity/lipophilicity will tend to migrate towards each other. Especially, if they are originally present in an environment that does not match closely with their lipophilicity. As mentioned previously, the difference in component activity provides the driving force for migration (Section 5.3). Migration can be minimized by:
- minimising the driving force, e.g. by not using a lipid-based barrier in a product that contains relatively high amounts of fats or oils, or by
- minimising the diffusivity. This can be achieved by, for example, using a bi-layer film, where a secondary, hydrophilic coating provides an additional barrier for lipophilic product components.

5.10.2. Making use of migration

It is also possible to make use of the migration behaviour of components. One can incorporate a wide diversity of components into edible coatings to achieve controlled release of these components.

Edible coatings can be applied with only this function in mind. It is possible to apply aqueous gels, such as starch, alginate or carrageenan gels for application in meat products, for example. In case release is fast, it is possible to design a layer with limited permeability.

Antioxidants

Oxidation processes often start at the product surface, through light activation, and/ or due to the availability of oxygen. It therefore makes sense to create a reservoir of antioxidants at the surface, which releases the antioxidants slowly into the product. In addition, even an ultraviolet quencher can be incorporated or chelating agents (e.g. EDTA) that reduce the risks of oxidation of components in the product. Finally, enzymes such as catalases may be incorporated to remove highly oxidizing components, such as peroxides.

Antimicrobial agents

Of course it is also possible to incorporate antimicrobial components in the film (Baldwin, 1994; Wong, 1994). In general, the use of antimicrobial agents in a film is more effective than application of the agents straight to the product surface. These can be used to prolong shelf life, but also to increase food safety. Organic acids such as benzoic, sorbic, lactic or propionic acid have been incorporated in coatings with success (Eswaranandam *et al.*, 2004). Natural or synthetic antibiotics such as nisin or imazalil can be introduced. Fungicides such as benomyl (especially for cheese, incorporated in cellulose-based or paraffin-based edible coatings) but also spice extracts, such as thymol or cumin and even enzymes have been used as antimicrobial agents (lysozyme). Other components studied include thiosulfinates, metal ions (silver), and parabens (Hotchkiss, 1995).

Other components

Any other component that one would like to add to a product can be incorporated in a film, usually with effects other than direct application into the product (and mostly less of such component will be needed). One can think here of (micro-)nutrients, but also of colorants or flavours. The highly concentrated form in the coating will strongly influence the colour appearance or the sensorial perception.

5.11. Guidelines for the selection of edible-barrier films

In the previous sections, various aspects of barrier formation are discussed. In this section, we summarize the most important aspects needed for basic calculations, and provide much needed experimentally determined parameters, through which the feasibility of a coating can be assessed.
1. *Thermodynamics.* If two compartments, each with different water activity a_w have to be separated by a barrier, it is first necessary to know what the initial a_w values are. Then, one needs to know at which a_w the product quality will be

affected, and which moisture level this implies (through a sorption isotherm). The difference between initial moisture level and moisture level at the onset of quality degradation gives the total amount of water that can be maximally transported. If you cannot find any data on this, you can also estimate what the final a_w value would be, when the moisture has been distributed completely. From this, one can find the total amount of water (in kg) that can be transported. As a rule of thumb, by taking 5 to 10% of this value, you get an estimate of the total amount of moisture migration that is permitted. To do this, one needs the sorption isotherm of both compartments. Then determine the driving force over the barrier in Pa (*i.e.* the difference in water vapour pressure at both sides of the barrier).

2. *Required time span.* Next, one needs to estimate the required shelf life, and the expected storage circumstances. It is prudent to make a pessimistic estimation here; best-case scenarios are expected to be unrealistic, also given the possibility of defects. High temperatures usually mean faster moisture migration, so one may estimate the temperature to be as high as realistically possible to obtain the worst case scenario.

3. *Estimation of the total permeance.* Determine the required permeance, by dividing the maximum amount of moisture migration by the required time span, and multiply by the interfacial area A for mass transfer between the two compartments and the driving force:

$$\Delta M = -M \cdot A \cdot t = \left(\frac{P}{L}\right)(p_0 - p_1) \cdot A \cdot t \Rightarrow permeance = \left(\frac{P}{L}\right) = \frac{\Delta M}{(p_0 - p_1) \cdot A \cdot t}$$

4. *Selection of a barrier material.* First estimate the typical thickness of the barrier you want to apply; in general, barriers of 100 μm and less are not visible to the naked eye. Determine the required maximal permeability ($P = permeance \cdot L$). Select a list of materials that would fulfil your requirements from the tables given in Section 5.12. If you cannot reach the required permeance, you might consider using a thicker film. However, make sure that you check the consequences for product perception. Also consider the use of composite materials (emulsions, dispersions, bi- and multi-layers) and the possibility of occurrence of defects.

5. *Consider the mechanics of coating and product.* Evaluate the consequences of thermal expansion and contraction. Is the product subject to temperature variations during further processing, or later in the supply chain? Is the product soft or elastic? Formulate the required mechanical properties of the coating, and then look to see whether this material has already been described. If not, consider the use of plasticizers, but be aware that a plasticizer will increase the water permeability significantly. It is difficult to predict the effect of plasticizers on water permeability; therefore try to look for analogues in the available data.

6. *Select an application procedure.* Consider the nature and morphology of the substrates (surface properties, porosity, mechanical properties); also evaluate what conditions the film will be subject to during production, packaging, storage, handling, retail and consumption. Is it possible to use a melt-based process? If not, can you use a solution or emulsion-based process? If you have chosen multiple coating, how will you apply each layer?

7. *Impact on product and process.* Consider the total costs of the system. Re-evaluate whether the coating fulfils the original requirements, and how it fits in the total product design and process system. Is there any consequence for shelf life? Are there any extra microbiological aspects to consider? Is there a chance that product integrity will be compromised if the coating fails, and if so, how can you avoid this?

These guidelines are by no means complete or universal. In fact, they suggest an iterative procedure, which can be used until a satisfactory combination of coating, product and process has been found.

The various aspects of coating technology discussed in this chapter offer a route to the rational design of a good coating. This rational design of the coating is one of the paths that lead to integrated product and process design.

5.12. Some experimental data

Some data are given below for the application of edible films as moisture barriers. These and other edible films are also used as barriers against oxygen or CO_2. It should be noted that all theory given above is valid, independent of the type of permeant. Therefore, the given background is by no means limited to the application as moisture barriers; it is only the data that is kept limited to moisture as a permeant, for reasons of compactness.

5.12.1. Sorption data

Material	Model	w_0	C	K	Conditions	References[1]; comments
Popcorn	BET*	4.843	28.31		10 °C	Maskan, 1999; w_0
	GAB	6.744	11.08	0.880	10 °C	is on weight% dry
	BET*	4.141	73.46		20 °C	basis, *for $a_w < 0.43$
	GAB	6.101	15.97	0.856	20 °C	
	BET*	3.300	97.30		30 °C	
	GAB	5.764	7.424	0.850	30 °C	
Popcorn coated	BET	4.672	5.023		10 °C	Maskan, 1999; w_0
with sugars	GAB	33.25	0.284	0.724	10 °C	is on weight% dry
	BET	3.955	8.971		20 °C	basis
	GAB	28.47	0.276	0.720	20 °C	
	BET	2.764	23.06		30 °C	
	GAB	24.01	0.254	0.744	30 °C	
Biscuit from	BET	6.53	19.88		14 °C	Arogba, 2001; w_0 is
wheat and		5.67	18.51		31 °C	on weight% fat-free
processed mango		4.48	23.52		58 °C	dry basis
kernels (1:1)						
Cracker	GAB	0.050	9.012	0.957	20 °C	Kim, 1998; w_0 in
		0.044	5.051	0.974	30 °C	kg water/kg solids;
		0.040	3.378	0.987	40 °C	strawberry jam:
Cookie		0.034	5.866	0.982	20 °C	37.3 wt% moisture;
		0.032	4.204	0.997	30 °C	thickness of cracker:
		0.030	2.979	1.010	40 °C	3 mm; thickness
Strawberry jam		0.135	75.773	0.990	20 °C	of cookie: 4 mm;
		0.080	42.034	1.009	30 °C	thickness of jam
		0.050	25.980	1.031	40 °C	layer: 2 mm
Cracker +		0.060	67.696	0.972	20 °C	
strawberry jam +		0.052	38.647	0.976	30 °C	
cracker		0.044	20.085	0.981	40 °C	
Cookie +		0.045	57.978	0.953	20 °C	
strawberry jam +		0.038	27.254	0.992	30 °C	
cookie		0.034	10.018	1.022	40 °C	
Minced meat	GAB	$-0.054T$ $+22.54$	exp(1279.3/T -1.966) exp($-4740/T$ $+17.9$)	exp(0.0012T -0.377)	< 40 °C > 40 °C	Lind, 1991; $T =$ temperature (K)

[1] Only the first author of the reference is given.

Remko Boom, Karin Schroën and Marian Vermuë

5.12.2. General permeability data

Test conditions: humidity given for both sides of the film during testing. Not sp. = Not specified

Coating material	Permeability for water vapour 10^{-13} kg/m·s·Pa	Diffusivity for water vapour 10^{-10} m²/s	Film thickness (μm)	Test conditions	References[1]
Gluten with glycerine as plasticizer	6.10		not sp.	not sp.	Saravacos, 2001
Gluten	5.00	1.00	not sp.	not sp.	Saravacos, 2001
Whey protein - sorbitol	7.20		not sp.	not sp.	McHugh, 1994
Corn pericarp	1.60	0.10	not sp.	not sp.	Saravacos 2001
Zein-glycerin	1.00	not sp.	not sp.	not sp.	McHugh, 1994
Sodium caseinate	4.26	not sp.	not sp.	not sp.	McHugh, 1994 , Avena-Bustillos, 1993
Sodium caseinate – glycerine (2:1)	35.91	not sp.	109	23 °C, 55-76% RH	Banerjee, 1994
Sodium caseinate – glycerine (2:1)	22.03	not sp.	105	23 °C,55-71% RH	Banerjee, 1994
	33.76	not sp.	111	23 °C, 55-76% RH	
Sodium caseinate – acetylated monoglyceride (1:1)	2.56	not sp.	88	25 °C, 0/88% RH	Avena-Bustillos, 1993
Sodium caseinate – acetylated monoglyceride- glycerine (2:2:1)	32.39	not sp.	141	23 °C, 55/72% RH	Banerjee, 19940
Calcium caseinate	3.26	not sp.	820	25 °C, 0-85% RH	Avena-Bustillos, 1993
Calcium caseinate – glycerine (2:1)	22.03	not sp.	105	23 °C, 55-71% RH	Banerjee, 1994
Calcium caseinate – acetylated monoglyceride – glycerine (2:2:1)	20.76	not sp.	157	23 °C, 55-67% RH	Banerjee, 1994
Calcium caseinate – beeswax (2:2:1)	0.95	not sp.	820	25 °C, 0/95% RH	Avena-Bustillos, 1993
Chocolate	0.1	not sp.	not sp.	not sp.	Saravacos, 2001
Beeswax	0.006	not sp.	not sp.	not sp.	McHugh, 1994
LDPE (for comparison)	0.0140	0.019	not sp.	not sp.	McHugh, 1994
Protein films, general	0.10 – 10.0	0.100	not sp.	not sp.	Saravacos, 2001
Polysaccharide films	0.10 – 10.0	0.100	not sp.	not sp.	
Lipid films	0.003 – 0.100	0.100·10	not sp.	not sp.	
Whey protein isolate – glycerine (1.6:1)	0.77	not sp.	110	25 °C, 0-11% RH	Krochta, 1994
	13.9		112	25 °C, 0-65% RH	

142   Food product design

Coating material	Permeability for water vapour 10^{-13} kg/m·s·Pa	Diffusivity for water vapour 10^{-10} m²/s	Film thickness (μm)	Test conditions	References[1]
Whey protein isolate – glycerine (4:1)	8.14	not sp.	130	25 °C, 0-77 % RH	Krochta,, 1994
Whey protein isolate – sorbitol (1.6:1)	7.19	not sp.	130	25 °C, 0-79 % RH	Krochta, 1994
Whey protein isolate – beeswax – sorbitol (3.5:1.8:1)	3.26 6.1	not sp.	190 140	25 °C, 0/93% RH 25 °C, 0/98% RH	McHugh, 1994
Lactic acid precipitated casein + sorbitol (0.6:1)	5.20	not sp.	203	37.8 °C, 0/90% RH	Chick, 1998
Lactic acid precipitated casein + sorbitol (1:1)	5.21	not sp.			
Lactic acid precipitated casein + sorbitol (1.4:1)	3.94	not sp.			
Lactic acid precipitated casein + glycerol (0.6:1)	6.35	not sp.			
Lactic acid precipitated casein + glycerol (1:1)	6.86	not sp.			
Lactic acid precipitated casein + glycerol (1.4:1)	6.33	not sp.	203	37.8 °C, 0/90% RH	Chick, 1998
Rennet casein + sorbitol (0.6:1)	5.75	not sp.			
Rennet casein + sorbitol (1:1)	5.74	not sp.			
Rennet casein + sorbitol (1.4:1)	4.58	not sp.			
Rennet casein + glycerol (0.6:1)	6.70	not sp.			
Rennet casein + glycerol (1:1)	6.74	not sp.			
Rennet casein + glycerol (1.4:1)	5.23	not sp.			

[1] Only the first author of the reference is given.

Remko Boom, Karin Schroën and Marian Vermuë

5.12.3. Lipid-based films

Coating material	Permeability for water vapour 10⁻¹³ kg/m·s·Pa	Diffusivity for water vapour 10⁻¹⁰ m²/s	Film thickness (μm)	Test conditions	References[1]
Paraffin wax	0.002	not sp.	not sp.	25 °C, 0/100% RH	Lovegren, 1954
Candelilla wax	0.002	not sp.	not sp.	25 °C, 0/100% RH	Greener-Donhowe, 1994
Carnauba wax + shellac	0.018	not sp.	not sp.	30 °C, 0/92% RH	Hagenmaier, 1992
Carnauba wax – glycerol monostearate	3.5	not sp.	not sp.	25 °C, 22/100% RH	Hugon, 1998
Carnauba wax – glycerol monostearate	0.136	not sp.	not sp.	25 °C, 22/65% RH	Hugon, 1998
Microcrystalline wax	0.003	not sp.	not sp.	25 °C, 0/100% RH	Greener-Donhowe, 1994
Beeswax	0.006	not sp.	not sp.	25 °C, 0/100% RH	Greener-Donhowe, 1994
Capric acid	0.038	not sp.	not sp.	23 °C, 12/56% RH	Koelsch, 1992
Myristic acid	0.347	not sp.	not sp.	23 °C, 12/56% RH	Koelsch, 1992
Palmitic acid	0.065	not sp.	not sp.	23 °C, 12/56% RH	Koelsch, 1992
Stearic acid	0.022	not sp.	not sp.	23 °C, 12/56% RH	Koelsch, 1992
Acetyl acyl glycerols	2.2 – 14.8	not sp.	not sp.	25 °C, 0/100% RH	Lovegren, 1954
Shellac	0.036 – 0.077	not sp.	not sp.	30 °C, 0/84% RH	Hagenmaier, 1992
Shellac	0.042 – 0.103	not sp.	not sp.	30 °C, 0/100% RH	Hagenmaier, 1992
Dewaxed gumlac	0.237	not sp.	not sp.	25 °C, 22/100% RH	Hugon, 1998
Dewaxed gumlac	0.151	not sp.	not sp.	25 °C, 22/75% RH	Hugon, 1998
Triolein	1.21	not sp.	not sp.	25 °C, 22/84% RH	Gallo, 1999
Hydrogenated cottonseed oil	0.013	not sp.	not sp.	26.7 °C, 0/110% RH	Guilbert, 1998
Hydrogenated palm oil	22.7	not sp.	not sp.	25 °C, 0/85% RH	Hugon, 1998
Hydrogenated peanut oil	39.0	not sp.	not sp.	25 °C, 0/100% RH	Lovegren, 1954
Native peanut oil	1.38	not sp.	not sp.	25 °C, 22/44% RH	Martin-Polo, 1991
Cocoa butter equivalent	0.06	not sp.	not sp.	25 °C, 22/84% RH	Gallo, 1999
Non-tempered cocoa butter	2.35	not sp.	not sp.	26.7 °C, 0/75% RH	Landmann, 1960
Tempered cocoa butter	0.49	not sp.	not sp.	26.7 °C, 0/75% RH	Landmann, 1960
Tempered cocoa butter	2.68	not sp.	not sp.	26.7 °C, 0/100% RH	Landmann, 1960
Milk chocolate	0.112	not sp.	not sp.	26.7 °C, 22/75% RH	Landmann, 1960
Milk chocolate	8.92	not sp.	not sp.	26.7 °C, 0/100% RH	Landmann, 1960
Dark chocolate	0.024	not sp.	not sp.	20 °C, 0/81% RH	Biquet, 1988

[1] Only the first author of the reference is given.

5.12.4. Bilayer films

Coating material	Permeability for water vapour 10^{-13} kg/m·s·Pa	Diffusivity for water vapour 10^{-10} m²/s	Film thickness (μm)	Test conditions	References[1]
Methyl cellulose – paraffin wax	0.02 – 0.04	not sp.	not sp.	25 °C, 22/84% RH	Martin-Polo, 1992
Methyl cellulose – paraffin oil	0.24	not sp.	not sp.	25 °C, 22/84% RH	Gallo, 1999
Methyl cellulose – beeswax	0.0058	not sp.	not sp.	25 °C, 0/100% RH	Greener-Donhowe, 1994
Methyl cellulose – carnauba wax	0.0033	not sp.	not sp.	25 °C, 0/100% RH	Greener-Donhowe, 1994
Methyl cellulose – candelilla wax	0.0018	not sp.	not sp.	25 °C, 0/100% RH	Greener-Donhowe, 1994
Methyl cellulose – triolein	0.076	not sp.	not sp.	25 °C, 22/84% RH	Gallo, 1999
Methyl cellulose – hydrogenated palm oil	0.049	not sp.	not sp.	25 °C, 22/84% RH	Gallo, 1999
Hydroxypropylmethyl cellulose – stearic acid	0.012	not sp.	not sp.	27 °C, 0/97% RH	Hagenmaier, 1990
Carnauba wax – glycerol monostearate – dewaxed gumlac	0.429	not sp.	not sp.	25 °C, 22/75% RH	Hugon, 1998
Carnauba wax – glycerol monostearate – dewaxed gumlac	0.069	not sp.	not sp.	25 °C, 22/65% RH	Hugon, 1998

[1] Only the first author of the reference is given.

5.12.5. Emulsion-based coatings

Coating material	Permeability for water vapour 10^{-13} kg/m·s·Pa	Diffusivity for water vapour 10^{-10} m²/s	Film thickness (μm)	Test conditions	References[1]
Methyl cellulose – polyethylene glycol 400 + behenic acid	0.77	not sp.	not sp.	23 °C, 12/56% RH	Kester, 1989
Methyl cellulose – triolein	1.44	not sp.	not sp.	25 °C, 22/84% RH	Gallo, 1999
Methyl cellulose – hydrogenated palm oil	1.32	not sp.	not sp.	25 °C, 22/84% RH	Gallo, 1999
Methyl cellulose – polyethylene glycol 400 – myristic acid	0.35	not sp.	not sp.	23 °C, 12/56% RH	Koelsch, 1992
Carboxymethyl cellulose – butter oil – pectin	1.51	not sp.	not sp.	25 °C, 22/84% RH	Debeaufort, 1994
Hydroxypropyl cellulose – polyethylene glycol 400 – acetylated monoglyceride	0.82	not sp.	not sp.	21 °C, 0/85% RH	Park, 1990
Wheat gluten – acetylated monoglyceride	0.56 – 0.66	not sp.	not sp.	23 °C, 0/11% RH	Gennadios, 1993
Wheat gluten – oleic acid	0.79	not sp.	not sp.	30 °C, 0/100% RH	Gontard, 1994
Wheat gluten – soy lecithin	1.05	not sp.	not sp.	30 °C, 0/100% RH	Gontard, 1994
Wheat gluten – paraffin wax	0.17	not sp.	not sp.	25 °C, 22/84% RH	Gallo, 1999
Wheat gluten – paraffin oil	0.51	not sp.	not sp.	25 °C, 22/84% RH	Gallo, 1999
Wheat gluten – triolein	0.97	not sp.	not sp.	25 °C, 22/84% RH	Gallo, 1999
Wheat gluten – hydrogenated palm oil	0.74	not sp.	not sp.	25 °C, 22/84% RH	Gallo, 1999
Sodium caseinate – acetylated monoglyceride	1.83 – 4.25	not sp.	not sp.	25 °C, 0/100% RH	Avena Bustillos, 1993
Sodium caseinate – lauric acid	1.1	not sp.	not sp.	25 °C, 0/92% RH	Avena Bustillos, 1993
Sodium caseinate – beeswax	1.11 – 4.25	not sp.	not sp.	25 °C, 0/100% RH	Avena Bustillos, 1993
Whey protein isolate – palmitic acid	2.22	not sp.	not sp.	25 °C, 0/90% RH	McHugh, 1994
Whey protein isolate stearyl alcohol	5.36	not sp.	not sp.	25 °C, 0/86% RH	McHugh, 1994
Whey protein isolate – beeswax	2.39 – 4.78	not sp.	not sp.	25 °C, 0/90% RH	McHugh, 1994

[1] Only the first author of the reference is given.

5.12.6. Mechanical properties of some edible films

	Tensile strength MPa	Elongation at breakpoint %	References[1]
α_{s1}-casein – glycerol (50:1)	$4.0 \cdot 10^{-4}$	38.0	Morr, 1993
α_{s1}-casein – glycerol (50:1) (transglutaminase treated)	$10.2 \cdot 10^{-4}$	74.3	Chen, 1993
Sodium caseinate – glycerol (2:1)	36.9	18.0	Banerjee, 1994
Sodium caseinate – glycerine (2:1) (transglutaminase treated)	2.98	29.9	Banerjee, 1994
Potassium caseinate (2:1) – glycerine (2:1)	2.97	42.8	Banerjee, 1994
Calcium caseinate – glycerine (2:1)	4.25	1.4	Banerjee, 1994
Whey protein concentrate – glycerine (2:1)	3.49	20.8	Banerjee, 1994
Whey protein isolate – glycerine (2:1)	5.76	22.7	McHugh, 1994
Whey protein isolate – glycerine (2.3:1)	13.9	30.8	McHugh, 1994
Whey protein isolate – glycerine (5.7:1)	29.1	4.1	McHugh, 1994
Whey protein isolate – sorbitol (2:1)	14.0	1.6	Banerjee, 1994
Sodium caseinate – glycerine – acetylated monoglyceride (2:1:2)	1.32	27.4	Banerjee, 1994
Potassium caseinate – glycerine – acetylated monoglyceride (2:1:2)	1.66	17.7	Banerjee, 1994
Calcium caseinate – glycerine – acetylated monoglyceride (2:1:2)	2.14	13.4	Banerjee, 1994
Whey protein concentrate – glycerine – acetylated monoglyceride (2:1:2)	1.08	13.6	Banerjee, 1994
Whey protein isolate – glycerine – acetylated monoglyceride (2:1:2)	3.14	10.8	Salame, 1986
Lactic acid precipitated casein + sorbitol (0.6:1)		170.7	Chick, 1998
Lactic acid precipitated casein + sorbitol (1:1)		156.0	
Lactic acid precipitated casein + sorbitol (1.4:1)		50.6	
Lactic acid precipitated casein + glycerol (0.6:1)		121.4	
Lactic acid precipitated casein + glycerol (1:1)		253.6	
Lactic acid precipitated casein + glycerol (1.4:1)		194.1	
Rennet casein + sorbitol (0.6:1)		4.9	
Rennet casein + sorbitol (1:1)		7.8	
Rennet casein + sorbitol (1.4:1)		17.9	
Rennet casein + glycerol (0.6:1)		123.2	
Rennet casein + glycerol (1:1)		185.4	
Rennet casein + glycerol (1.4:1)		223.5	
LDPE (for comparison)	13 – 32	36 – 500	Salame, 1986, Chen, 1995

[1] Only the first author of the reference is given.

References

Arogba, S.S. (2001). Effect of temperature on the moisture sorption isotherm of a biscuit containing processed mango (Mangiflora indica) kernal flour. J. Food Eng. 48: 121-125.

Avena-Bustillos, R.J. and Krochta, J.M. (1993). Water vapour permeability of caseinate-based edible films as affected by pH, calcium cross-linking and lipid content. J. Food Sci. 58: 904-907.

Baldwin, E.A. (1994). Edible coatings for fresh fruits and vegetables: past present and future. In: Edible coatings and films to improve food quality. J.M. Krochta et al.(eds.), Technomic Publ. Co., Lancaster, PA., USA, p25-64.

Banerjee, R. and Chen, H. (1994). Functional properties of edible films using whey protein concentrate. J. Dairy Sci. 78: 1673-1683.

Biquet, B. and Labuza, T.P. (1988). 'Evaluation of the moisture permeability characteristics of chocolate films as an edible moitsure barrier.' J. Food Sci. 53 (4): 989-997.

Bruin, S. and Luyben, K. (1980). Drying of food materials: a review of recent developments. In: Advances in drying. A. S. Mujumbar (ed). Hemisphere, N.Y. USA, p155.

Brunauer, S., Emmett, P.H. and Teller, E., (1938). Adsorption of gases in multimolecular layers. J.Am. Chem.Soc. 60: 309-314.

Chen, H., Banerjee, R. and Wu , J. (1993). Strengths of thin films derived from whey proteins. Paper nr 93-6528. Am. Soc. Agric. Eng, St. Joseph. MI, USA.

Chen, H. and Zhang S. (1994). Effects of protein-lipid and protein-plasticizer ratios on functional properties of sodium caseinate-acetylated monoglyceride films. J. Dairy Sci. 77 (Suppl. 1): 8.

Chen, H. (1995). Functional properties and applications of edible films made of milk proteins. J. Dairy Sci. 78: 2563-2583.

Chick, J. and Ustunol, Z. (1998). Mechanical and barrier properties of lactic acid and rennet precipitated casein-based edible films. J. of Food Sci. 63(6): 1024-1027.

Debeaufort, F. (1994). Etude des transferts de matière au travers de films d'emballage - perméation de l'eau et de substances d'arôme en relationi avec des propriétés physico-chimiques des films comestibles. Dijon, Université de Bourgogne, France.

Eswaranandam, S., Hettiarachchy, N.S. and Johnson, M.G. (2004) Antimicrobial Activity of Citric, Lactic, Malic, or Tartaric Acids and Nisin-incorporated Soy Protein Film Against Listeria monocytogenes, Escherichia coli O157:H7, and Salmonella gaminara. J. Food Science 69 (3): 79-84.

Gennadios, A., Weller, C.L. and Testin, R.F. (1993). 'Property modification of edible wheat gluten-based films.' Trans. ASAE 36: 465-470.

Gmehling, J. and Onken. U. (1977). Vapor-Liquid Equilibrium Data Collection. DECHEMA Chemistry Data Series. Frankfurt, 1 (parts 1-10).

Gontard, N., Duchez, C., Cuq, J.-L. and Guilbert, S. (1994). Edible composite films of wheat gluten and lipids: water vapour permeability and other physico-chemical properties. Int. J. of Food Sci. Technol. 29: 39-50.

Greener-Donhowe, I. and Fennema, O. (1994). Edible films and coatings: characteristics, formation, definitions and testing methods. pp. 1-24 In: J.M. Krochta et al. (eds.). Edible coatings and films to improve food quality. Technomic Publ. Co., Lancaster, PA., USA.

Guilbert, S. and Cuq, B. (1998). Les filmes et enrobages comestibles. In: L'emballage des denrées alimentaires de grande consommation. G. Bureau, J.A. Multon (eds). Paris: Tec & Doc Lavoisier, APRIA., France, pp 320-359.

Hagenmaier, R.D. and Shaw, P.E. (1990). Moisture permeability of edible films made with fatty acid and (hydroxypropyl)methylcellulose. J. Agric. Food Chem. 38: 1799-1803.

Hagenmaier, R.D. and Shaw, P.E. (1992). Gas permeability of fruit coating waxes. J. Am. Soc. Hort. Sci. 117: 105-109.

Hotchkiss, J.H. (1995). Safety considerations in active packaging. pp 238-255 In: Active food packaging. M. L. Rooney (ed.) Blackie Academic & Professional, Glasgow, United Kingdom.

Hugon, F. (1998). Etude et maîtrise des transferts d'eau dans des céréales enrobées. D.R.T., ENSBANA, Université de Bourgogne, Dijon, France.

Kester, J.J. and Fennema, O. (1989). Resistance of lipid films to water vapour transmission. JAOCS 66: 1139-1146.

Kim, S.S., Kim, S.Y., Kim, D.W., Shin, S.G. and Chang, K.S. (1998). 'Moisture sorption characteristics of composite foods filled with strawberry jam.' Lebensm.-Wiss. u.-Technol. 31: 397-401.

Koelsch, C.M. and Labuza, T.P. (1992). 'Functional, physical and morphological properties of methyl cellulose and fatty acid-based edible barriers.' Lebensm.-Wiss. u.-Technol. 25(5):404-411.

Krochta, J.M., Baldwin, E.A. and Nisperos-Carriedo, M.O. (1994). Edible Coatings and films to improve food quality. Lancaster, PA: Technomic Publ. Inc., USA.

Landmann, W. Lovegren, N.V. and Feuge, R.O. (1960). Permeability of some fat products to moisture. JAOCS 37: 1-4.

Lind, I. and Rask, C. (1991). Sorption isotherms of mixed minced meat, dough and bread crust. J. Food Eng. 14: 303-315.

Lovegren, N.V. and Feuge, R.O. (1954). Food coatings. Permeability of acetostearin products to water vapour. J. Agric. Food Chem. 2: 558-563.

Martin-Polo, M., Mauguin, C. and Voilley, A. (1992). Hydrophobic films and their efficiency against moisture transfer. I. Influence of the film preparation technique. J. Agric. Food Chem. 40: 407-412.

Martin Polo, M. (1991). Influence de la nature et de la structure de films et d'enrobages alimentaires sur le transfert de vapeur d'eau. Dijon, Université de Bourgogne, France.

Maskan, M. and Gögüs, F. (1999). Water adsorption properties of coated and non-coated popcorns. J. of Food Proc. Preservation 23: 499-513.

McHugh, T.H. and Krochta, J.M. (1994). Dispersed phase particle size effects on the water vapour permeability of whey protein beeswax edible emulsion films. J. Food Process Preserv. 18: 173-188.

McHugh, T.H. and Krochta, J.M. (1994). Permeability properties of edible films. In: Edible films and coatings to improve food quality. J.M. Krochta et al. (eds.). Lancaster, Pa.: Technomic Publishing Co. USA, pp 139-187.

McHugh, T.H. and Krochta, J.M. (1994). Sorbitol- vs. glycerol-plasticized whey protein edible films: integrated oxygen permeability and tensile property evaluation. J. Agric. Food Chem. 42: 841-845.

Morr, C.V. and Ha, E.Y.V. (1993). Whey protein concentrates and isolates: processing and functional properties. Crit. Rev. Food Sci. Nutr. 33: 431-476.

Quezada-Gallo, J.A. (1999). Influence de la structure et de la composition de réseaux macromoléculaires sur les transferts de molécules volatiles (eau et arômes). Application aux emballages comestibles et plastiques. Dijon, Université de Bourgogne, France.

Park, H.J. and Chinnan, M.S. (1990). Properties of edible coatings for fruits and vegetables. Am. Soc. Agric. Engin., Paper N° 90-6510, Chicago, IL, USA.

Poncelet, D. (2006) Microencapsulation:fundamentals, methods and applications In: Surface Chemistry in Biomedical and Environmental Science. J.P. Blitz and V.M. Gun'ko (eds.) Springer, the Netherlands, pp. 23-34.

Salame, M.(1986). Barier Polymers, In: The Wiley encyclopaedia of packaging technology. M. Bakker (ed), J. Wiley & Sons, New York, USA, pp 46-54.

Saravacos, G.D. and Maroulis, Z.B. (2001). Transport properties of foods. New York, Marcel Dekker, Inc.

Van Boekel, A.J.S. (2008) Kinetic Modeling Of Reactions In Foods, CRC Press., Boca Ration, FL, USA.

Weiser, H (1985) Influence of temperature on sorption equilibria. In: D. Simatos and J.L Multon (eds.) Properties of Water in Foods. Martinus Nijhoff Publishers, Dordrecht, The Netherlands, pp 95-117.

Wong, D.S., Camirand, W.M. and Pavlath, A.E. (1994). Development of edible coatings for minimally processed fruits and vegetables. In: Edible coatings and films to improve food quality. J.M. Krochta et al.(eds.). Technomic Publ. Co, Lancaster, PA., USA, pp 65-82.

6. Emulsions: properties and preparation methods

Karin Schroën and Remko Boom
Laboratory of Food Process Engineering, Wageningen University

6.1. Introduction

Emulsions are dispersions of immiscible fluids, e.g. oil and water. In many industries, these liquids need to be mixed intimately in order to obtain a fine emulsion. Some examples of emulsions are paints, spreads, sauces, cosmetic crèmes, pharmaceutical ointments, etc. Since a (macro) emulsion is intrinsically unstable, it needs to be stabilized by components (small molecules such as surfactants, larger polymers, or even particles can be used to form so-called Pickering emulsions) that adhere to the formed interface, which may also reduce the interfacial tension, therewith reducing one of the driving force for coalescence.

Mostly the emulsification process is started with a mixture of oil and water that is already crudely mixed into a coarse emulsion. Upon passage of the emulsification device, the coarse emulsion is broken up into smaller droplets, and quite often, this process needs to be repeated to obtain the desired droplet size with a sufficiently narrow distribution in droplet size. The methods of choice for industrial emulsification are high-pressure homogenization, rotor-stator systems, and ultrasound treatment (see Figure 6.7 for a schematic representation).

During the creation of the oil/water interface, which mostly takes place in a highly turbulent environment in the previously mentioned methods, the surface needs to be stabilized sufficiently fast in order to prevent coalescence right after formation. For this, it is essential that the action of the stabilizer is tuned to the creation process. However, this type of information is mostly not available. Therefore, emulsification often seems more of an art than a science, also due to the very short time scales at which things occur, which requires high-speed recording in a highly turbulent environment. Obviously, that is not easy to do; in some cases, emulsification can be simplified and investigated with microfluidic devices, as presented in the last part of this chapter.

Still, considerable knowledge about emulsions, emulsification, its energy requirements, and application to various related products such as encapsulates, has been reported. In this chapter, we will present examples of food emulsions and their properties including surfactant/stabilizer behaviour in Section 6.2, the basics of emulsification methods currently used in the industry will be discussed in Section 6.3, including guidelines that can be used in the design of emulsification processes. Furthermore, we will show the newest developments in emulsification technology in Section 6.4, and compare them to the current emulsification techniques. In the last section of this chapter, we will give an outlook on developments in emulsification science, and look into the future.

The interested reader can be referred to the Encyclopedia of Emulsion Technology edited by Becher, or to the books written by Walstra (2003, and with co-workers, 2006) for extensive descriptions of the current emulsification technology.

6.2. Emulsions

6.2.1. Food emulsion products

In an emulsion, usually one fluid is present as small droplets in another phase. The basic forms are oil in water (abbreviated as O/W), and water in oil (W/O) emulsions. The droplet phase is called the dispersed phase (prior to emulsification, we will use the term to-be dispersed phase), the surrounding phase the continuous phase. Emulsions can be found in many different products. Some examples are:

- Salad dressings made by emulsifying vegetable oil in an aqueous mixture that contains acid (e.g. vinegar or lemon) and other taste components such as mustard. When made at home, this emulsion is rather unstable because mustard is not a very efficient emulsifier: the droplets coalesce relatively quickly so one has to shake it before use. Commercial variants are usually stabilized by other components.
- Mayonnaise is a very concentrated emulsion of oil droplets in acidified water (lemon works very well due to its low pK), stabilized by proteins from egg (yolk) that are maximally charged due to the citric acid from the lemon. Mayonnaise is very concentrated (70-80% v/v), and the droplets are squeezed together but do not coalesce due to the charge of the stabilizers, which gives the mayonnaise its nice consistency.
- The previously mentioned egg yolk is an emulsion of its own right, it consists of egg fat (and cholesterol) and protein in an aqueous solution, stabilized by a mixture of phospholipids.
- Processed milk is an emulsion of milk fat stabilized by phospholipids and various proteins in an aqueous solution containing many different proteins, lactose, and salts. In raw milk, the fat is present in the form of milk fat globules, which are surrounded by a membrane consisting of phospholipids and many proteins derived from the lactating cell. When this milk is homogenized in the factory, the globules are broken, and the fat is dispersed into smaller droplets. For stabilization, the material that is initially present in the interface is not sufficient, and components from the plasma such as proteins adsorb and act as stabilizers.
- Cream, including various culinary products, is a concentrated emulsion of milk fat in an aqueous phase; the concentration depends on the type of cream. Stabilization is as described for milk.
- Margarine is an emulsion of water droplets in fat, stabilized by a packing of needle-like crystals of fat inside the continuous fat phase. The same is also true

for butter, although it should be mentioned that this product is obtained through phase inversion.

Various other examples of food emulsions have been described in literature, for further reference see e.g. the Encyclopedia of Emulsion Technology (Becher, 1996), Krog (1985), or Walstra (2003, 2006).

Foams are closely related to emulsions, and sometimes they are used in tandem to give the emulsion a 'lighter' perception. In foams, the dispersed phase is not an oil or water, but is a gas. One can use similar techniques for making foam as for making emulsions, and some of the properties are comparable. Obviously, also foams are used often, foam on beer is well known (although not appreciated in all cultures), and the same is true for bread in which yeast is used to aerate the dough. Another well-known example is whipped cream, and a nice example of an aerated emulsion is ice cream, which consists of cream with a stabilizer (mostly gelatine) that captures the air bubbles that are incorporated in the liquid ice cream mix, that is further stabilized by crystal formation.

Besides single emulsions (O/W, and W/O), also double or duplex emulsions are reported, which are emulsions in emulsions. One can have water droplets in an oil phase, of which larger droplets have been made in a second aqueous phase: this is called a water-in-oil-in-water duplex emulsion (W/O/W). Of course, one can also have the reverse (O/W/O emulsion). W/O/W emulsions are used in some medical applications (encapsulation of drugs), and in foods, to enhance the perception of for example fat, but also to mask the taste of for example bitter peptides, and reduce the caloric load.

6.2.2. Emulsion properties

Any surface between two immiscible fluids has an interfacial energy (tension). This is caused by the difference in cohesion between molecules of the two phases. Creating an interface between two phases costs energy, proportional to the amount of interface generated. In practice much more than this minimal amount of energy needs to be invested due to energy loss to heat development (which may also damage temperature sensitive components, and through this the product properties), coalescence, and the necessity of having repeated passes through the emulsification device. The amount of energy that is needed to create 1 m^2 of interfacial area is called the interfacial energy (tension), with unit J/m^2 or N/m (for more information on surface phenomena we like to recommend e.g. Hiemenz, 1996 or Lyklema, 1991).

$$\Delta E = \int_0^A \sigma dA \tag{6.1}$$

in which σ is the interfacial energy/tension (N m^{-1} or J m^{-2}), A the total interfacial area created (m^2), and ΔE the energy needed to create this interface (J). The interfacial tension is always positive, with the exception of so-called micro-emulsions, which are considered outside the scope of this chapter. This means that ΔE is always positive, or put differently: making small droplets costs lots of energy, given the relation with the surface area that scales with the reciprocal value of the droplet size as shown in Equation 6.2 (Walstra, 2003).

$$\frac{A}{V} = \frac{3}{r_{drop}} = \frac{6}{d_{drop}} \qquad (6.2)$$

in which A is the available surface area (m²), V the volume of dispersed phase (m³), r_{drop} the droplet radius (m), and d_{drop} the droplet radius (m).

Since any system strives to minimize its energy, there is a driving force for droplet coalescence, which implies that emulsions are intrinsically unstable. As can be deduced from Equation 6.1, a reduction in the available interfacial area will yield a reduction in energy, and in order to prevent this, the emulsion needs to be protected. How this is done is discussed in the next section, but what it comes down to is, that a barrier needs to be created that prevents close contact of droplets.

In a spherical droplet, the interface is curved and the interfacial tension exerts a force perpendicular to the interface, directed to the concave side of the interface. The size of the force is reciprocally proportional to the curvature and can be described as a pressure; the so-called the Laplace pressure ($\Delta P_{laplace}$). In its most general shape, the Laplace pressure is given by:

$$\Delta P_{Laplace} = \sigma \left(\frac{1}{R_1} + \frac{1}{R_2} \right) \qquad (6.3)$$

with R_1 and R_2, the respective curvatures of the surface, and σ the interfacial tension (N m⁻¹). For a flat surface, the curvature is infinite, and $1/R$ becomes zero. For example in a cylinder, one curvature is given by the radius of the cylinder, while the other curvature does not contribute to the Laplace pressure (the length of the cylinder can be considered infinitely long compared to the radius of the cylinder). For any geometric shape, the resulting Laplace pressure can be found by simply adding up the two curvatures as shown in Equation 6.3. For a spherical droplet, of size R_d, both curvatures are equal and the resulting Laplace pressure is equal to:

$$\Delta P_{Laplace} = \frac{2\sigma}{R_d} \qquad (6.4)$$

From this it can be concluded that in droplets or other fluid domains surrounded by an interface, the pressure is always higher than that of the surroundings, and the smaller the curvature(s) of the domain, the higher the pressure difference (Walstra, 2003).

6.2.3. Droplet size and emulsion characterization

The size of the emulsion droplets determines the minimum amount of energy that is needed to form the emulsions (the stability against e.g. creaming is amongst others determined by the particle size; see 6.2.5). Depending on the size of the droplets, the emulsion may look transparent or have a certain colour. Very small droplets do not deflect the light and therefore the emulsions look transparent, while larger droplets act like little lenses that scatter the light. For monodisperse particles in the higher nm range, the liquid may have a specific colour, but polydisperse emulsions containing droplets in the range 50 nm-10 µm (like milk) always look milky white.

Clearly, the particle size is an important parameter, also concerning how much stabilizer is to be used (see 6.2.4); however, in practice the particle size is not easy to determine. Most emulsions are polydisperse, which implies they contain droplets of many different sizes; e.g. they show a droplet size distribution. In the following part, we describe droplet size distributions and characterization thereof.

Various methods can be used to characterize emulsions ranging from microscopic imaging and classification, which tends to be very laborious, to fully automated (dynamic) light scattering methods, such as the coulter counter, or the mastersizer. All methods have their pros and cons, and results always need to be interpreted with care; e.g. the automated methods use a curve fit procedure to smooth the results, and this may imply that the actual signal is rather different from the fitted curve.

Once data collection has taken place, the result may look as depicted in Figure 6.1, in which all counted droplets of a specific size are shown in Figure 6.1b, and the cumulative droplet size distribution in Figure 6.1a; the number of droplets on the y-axis is the percentage that is smaller than the value indicated on the x-axis.

There are different ways of deriving an average droplet size from the curve; the most general form is given in Equation 6.5. Obviously, one can count all droplets as one, and simply determine the average value; this is called the $d_{1,0}$ value (see also Equation

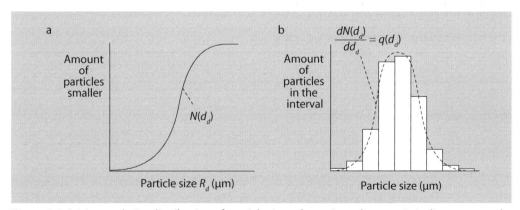

Figure 6.1. (a) A cumulative distribution of particle sizes; the various characteristic diameters can be determined directly from the values at the y-axis; and (b) a frequency distribution; the dashed line in the frequency distribution is the derivative of the cumulative distribution.

6.6). However, for emulsions that is not common practice; especially the large droplets are of great relevance for the emulsion stability, therefore, mostly the $d_{3,2}$ or sauter diameter (or even $d_{4,3}$) is used. This is a volume over surface area average as described in Equation 6.7, and in this way, the large volume of oil contained in the large droplets (scales with r^3) is given a bigger role. In contrast, in $d_{1,0}$ the small droplets, that are present most abundantly, are given most weight. All the different average diameters can be determined with Equation 6.5, using different values for m and n.

$$d_{nm} = \left(\frac{\sum_i d_i^n n_i}{\sum_i d_i^m n_i} \right)^{\frac{1}{n-m}} \tag{6.5}$$

In which d denotes diameter, and n number. For $d_{1,0}$ and $d_{3,2}$, the equation is as follows:

$$d_{10} = \frac{\sum_i d_i\, n_i}{\sum_i n_i} \; ; \qquad d_{32} = \frac{\sum_i d_i^3 n_i}{\sum_i d_i^2 n_i} \tag{6.6 and 6.7}$$

Please note that the values of these averages can be very different! Although it is convenient to characterize an emulsion through one average number, this also discards a lot of information, such as the width of the distribution. Especially the large particles can be of great influence on the stability of an emulsion. Therefore, also for the entire distribution a characteristic number is used, the so-called span. It is defined as:

$$Span = \frac{d(90) - d(10)}{d(50)} \tag{6.8}$$

where d(x) is the droplet size that has x % of droplets smaller than this value. The span is a value for the relative width of the distribution, and can be used for any of the distributions described earlier.

6.2.4. Surface active components

As mentioned previously, macro-emulsions are intrinsically instable, and in time, they will de-mix into their constituent liquids, and form two liquids on top of each other. Whether this will occur within a specific time span depends amongst others on the effectiveness of the added surfactant/stabilizer. Emulsion stability in general is discussed in 6.2.5, here we only highlight the surface active components. Three options are available:
1. *Surfactants or (bio)polymers* that adsorb to the interface, giving rise to electrostatic or steric repulsion between the droplets, and lower the interfacial tension; therewith reducing contact between droplets, and lowering the energy of the system, and the Laplace pressure.
2. *Polymers* that increase the viscosity of the continuous phase will reduce the droplet creaming velocity, and increase the shelf life of the product. Alternatively,

they may also give the product a yield stress, and if the strength of the weak gel is high enough, it can prevent creaming.

3. *Particles* that are partially wetted by both phases of the emulsion can accumulate in the interface, and give so-called Pickering stabilization, which is mostly a steric effect. *In situ* particle formation as is the case in margarine and butter can be considered a special case.

Alternatively, in very few cases, it is possible to match the densities of both liquids. In that case, the creaming velocity will be zero, but contact between droplets should still be prevented in order to prevent coalescence.

More information on surfactants and their behaviours can be found e.g. in Walstra, 2003; Walstra *et al.*, 2006, Lucassen-Reynders, 1996; Guzey and McClements, 2006).

Surfactants

Surfactant molecules are mostly low molecular weight components that have affinity for both constituent liquids of the emulsion, and therefore like to sit in the interface, with the polar part directed to the watery phase, and the apolar, or hydrophobic part to the oily phase. The hydrophobic part of the molecule mostly consists of an alkyl chain; the hydrophilic part can be a non-ionic chain, such as an ethylene oxide chain or a chain consisting of glucose units, or a chain containing charged groups (such as a carboxylic acid or a sulphonate group).

When the surfactants are adsorbed at the interface, they may form a barrier for two droplets to come very close through steric hindrance or electrostatic repulsion. When a non-ionic surfactant adsorbs on an interface, the hydrophobic part will be in the oil phase and the hydrophilic part will stick out into the aqueous phase. This gives the droplet a protective coat of hydrophilic chains that would make contact first before the oil droplets would be able to meet, and mostly would effectively form a steric hindrance for droplet-droplet contact. The only way that the two droplets could coalesce is when the non-ionic surfactant molecules would move away from the contact point; however, the molecules are strongly adsorbed, and cannot be removed easily. Hence, when two droplets with such a layer approach each other, the coats will repel each other; the droplets will move apart again, and coalescence is prevented. For charged surfactants, the droplets would have an effective charge (only in water continuous emulsions) that adds to the previously described effects, and makes that these droplets are efficiently protected against contact, and coalescence (Becher, 1996; Lucassen-Reynders, 1996; Walstra, 2003).

Because of the special structure of the surfactants, they are hard to dissolve molecularly in one of the two phases of an emulsion. Often, a surfactant will form micelles; for example in water, the hydrophobic parts are clustered together in the centre and the polar parts are on the outside; in oil reversed micelles are formed. In an emulsion, only a limited amount of surfactant can reside in the interface (typically in the order of 0.1-10 mg/m^2, depending on the surfactant/stabilizer). As soon as the surface is filled, the surplus of surfactant will remain in that phase for which it has most affinity. An excess of surfactant/stabilizer will be used in order to allow the surface to fill rapidly and therewith prevent coalescence. In literature, HLB (hydrophilic-lipophilic balance) values are used to predict the phase into which the surfactant will dissolve, but mostly this will be determined experimentally due to the complexity of the emulsions (Becher, 1996).

In general, the phase in which the surfactant can be dissolved becomes the continuous phase (Bancroft rule). This can be understood as follows: assume that two droplets approach each other, and the continuous phase is squeezed out (Figure 6.2a). While flowing out, the continuous phase exerts a force on the adsorbed surfactants, and the surfactant molecules flow along the surface that becomes locally depleted of surfactant. As a result, a concentration gradient of surfactant molecules on the surface is created, which counteracts the drag by the flow of the continuous phase, and reduces the outflow of the continuous phase and therewith the concentration gradient (Marangoni effect). If the surfactant is dissolved in the disperse phase (Figure 6.2b), the depleted surfactant will simply be replenished by the surfactant that is present in the droplet, and the situation will persist leading to further outflow of the continuous phase and close approach of the droplets (Walstra, 2003; Lucassen-Reynders, 1996). Besides, mostly the largest part of the molecule will determine the phase in which the molecule will dissolve, and since this is the largest part it will give thicker layers and better protection against coalescence.

Proteins

In most food products, proteins are used as stabilizers, and besides they act as surfactants that lower the interfacial tension, but not as much as the previously described surfactants. Proteins are polymer chains of amino acids; some amino acids are very hydrophilic (have a strong interaction with water), due to charged groups, others are more hydrophobic. Because of this ambivalence, proteins like to nest in interfaces, and are able to adjust their structure to some extent to optimize the interaction with both phases. The protein forms multiple interaction sites with the interface; therefore, its adsorption can be considered irreversible (all sites will not be

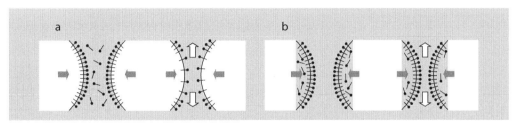

Figure 6.2. Schematic representation of the Bancroft rule: (a) surfactants in the continuous phase and (b) in the dispersed phase.

released simultaneously, this in contrast to low-molecular weight surfactants, which are at equilibrium with the surrounding phases) (Walstra *et al.*, 2006).

A protein is a large molecule; therefore, it diffuses slowly through the aqueous phase, towards the interface, and the typical time scales for surface formation need to be tuned to the diffusion behaviour if the protein should facilitate formation of new surface (i.e. lower interfacial tension). Mostly, these time scales do not match, which implies that protein is used at a relatively high concentration, or other measures are taken to facilitate droplet formation (e.g. add low dosage of surfactants).

Apart from an effect on the interfacial tension (which is mostly not big), proteins are very efficient stabilisers (see also emulsion stability section). The proteins form a layer on the surface that can act as a steric barrier for emulsion droplets that approach each other closely. A special case is casein, a large micelle (20-150 nm) consisting of sub-micelles, that either adsorbs as a whole, or spreads into sub-micelles. It is preferentially found in the interface after homogenisation, and this can be explained using the Kolmogorov theory (for more information see Walstra *et al.*, 2006). The main effect is that casein immediately covers a large area, therewith hindering other adsorbing species.

Non-adsorbing polymers

As described in 6.2.5, adding polymers to the continuous phase will increase its viscosity and will lead to low creaming/sedimentation velocity. Besides, it will also slow down the outflow of the continuous phase between approaching droplets, and through this slow down coalescence. An alternative option of using polymers is to create a weak gel (as is applied in e.g. chocolate milk), that traps the droplets. If the gel is sufficiently strong (i.e. has appropriate yield stress) it can resist the upward force exerted by the droplet, while it will still show flow behaviour. This is a very effective way for stabilizing an emulsion (Walstra, 2003), but cannot be applied universally.

Particles: pickering emulsions

Very small particles that are not completely wetted by either the continuous or the dispersed phase can be used to stabilize emulsions as depicted in Figure 6.3. The particles partially stick out of the interface, giving steric hindrance against coalescence, in a similar way as non-ionic surfactants do. The stabilization is better when the particles stick out further, i.e. they act better when they are wetted better by the continuous phase. Also here, a form of the Bancroft rule applies: the phase that wets the particles best will be the continuous phase (Walstra, 2003).

6.2.5. Emulsion stability

In the previous sections, we have hinted at various reasons why an emulsion may be unstable. In this section, we will use the term emulsion stability to describe

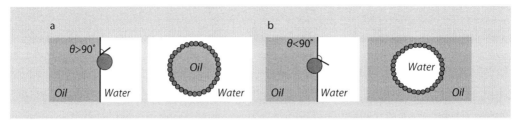

Figure 6.3. (a) Particles that are preferentially wetted by water, will mostly form an oil-in-water emulsion; (b) particles that are preferentially wetted by oil will form a water-in-oil emulsion.

aggregation of droplets, and coalescence of droplets. Various aspects that play a role mostly during preparation (bridging and depletion flocculation), and during storage (sedimentation) are discussed, and we are aware that we will be far from complete in our description, but try to stress the main issues that play a role. For background information, we like to suggest Walstra (1996).

Bridging and depletion flocculation

As mentioned in 6.2.4 on surface-active components, it is of great importance to cover the available surface to give it a protective coat. Even if this is the case, this does not necessarily mean the emulsion will be stable; this depends on the 'quality' of the coat. Various short-range interactions as described in the DLVO theory can allow droplets to approach and flocculate in a primary and secondary minimum (Lyklema, 1991), and these effects may become pronounced if multiple components (that e.g. carry opposite charges) are used to stabilize an emulsion. Describing these systems is complex, and still part of an on-going debate in literature. Here we will describe the main effects that can be used to establish a good starting point for emulsion preparation; whether this will also work in practice will always have to be tested.

There are two situations that will lead to flocculation, namely one in which the surface is not sufficiently covered, and a surface-active molecule can attach to two or even more droplets (Figure 6.4a). The other occurs if a non-adsorbing molecule is present; this is termed depletion flocculation (Figure 6.4b and c) (e.g. Walstra, 1996).

The component that causes depletion flocculation is non-adsorbing, which implies that the component cannot come as close to the surface as an adsorbing species would. For e.g. a polymer, it will always remain the distance of its gyration radius away from the interface, and because of that, there will be a lower polymer concentration in this area (depicted by the dotted line in Figure 6.4b and c). The concentration difference implies that there is also an osmotic pressure difference in the system; the system will strive to minimize this effect and can do so through minimization of the area that is free of polymer, i.e. aggregation. For spherical droplets this energy gain may be not very big (but still sufficient), but for platelet type of geometry this effect can be dominating.

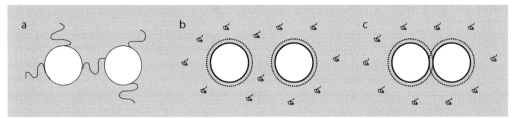

Figure 6.4. (a) A schematic representation of bridging flocculation; one molecule attaches to two or more droplets due to insufficient surface coverage. (b) and (c) show depletion flocculation by a non-adsorbing species. This species will have a non-assessable area near the droplets (indicated by the dotted line) that corresponds to the gyration radius of the component, and because of that, there will be a difference in osmotic pressure, which can be reduced by aggregation of droplets (see text for further explanation).

In the group of Julian McClements (University of Massachusetts, USA) these effects have been very practically translated into so-called stability maps that are very useful in choosing appropriate compositions for emulsions that are prepared with multiple components. In Figure 6.5a and b, we have added examples to show the practicality of such maps; many different ones are available, and we recommend using them as starting point for preparation of any emulsion. Figure 6.5a is a stability map showing the influence of droplet concentration on the critical polyelectrolyte concentrations for saturation, depletion, and adsorption. It was assumed that the droplets had a radius of 0.3 µm, and the non-adsorbed polyelectrolyte had a molecular weight of 100 kDa and effective radius of 30 nm. The shaded area highlights the range of conditions where it should be possible to produce non-flocculated droplets. Figure 6.5b is a stability map showing the influence of droplet size on the critical polyelectrolyte

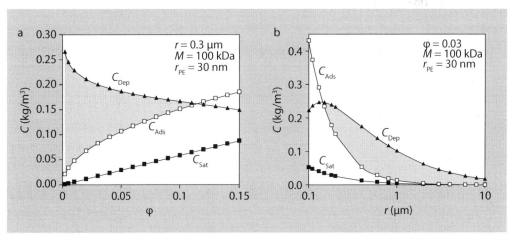

Figure 6.5. Stability map showing the influence of (a) droplet concentration and (b) droplet size on the critical polyelectrolyte concentrations for saturation, depletion, and adsorption. (Reprinted with permission by Elsevier from Guzey and McClements (2006).

concentrations for saturation, depletion, and adsorption. It was assumed that the droplets had a volume fraction of 0.03 (3 vol.%), and the non-adsorbed polyelectrolyte had a molecular weight of 100 kDa and effective radius of 30 nm. The shaded area highlights the range of conditions where it should be possible to produce non-flocculated droplets.

Sedimentation

Besides aggregation, also other physical phenomena will influence the stability of emulsions. The density difference between the two phases that normally is present will make droplets want to cream or sediment at a certain rate, depending on the density difference, which co-determines the stability of emulsions that are free flowing. The train of thought described in this section holds for free flowing emulsions, and not for (partially) crystallized emulsions or gelled emulsions that exhibit a high enough yield stress to keep the droplets captured (Section on polymer). To keep things simple, we will use the term sedimentation only, but obviously, the same principle holds for creaming.

When considering a droplet in an infinite amount of liquid, it is subjected to a buoyancy force due to the density difference, and a friction force exerted by the continuous phase. For gravity conditions, this so-called Stokes Law can be used to describe the sedimentation velocity as follows:

$$v_s = \frac{g \cdot \Delta \rho \cdot d_{drop}^2}{18 \cdot \eta_c} \qquad (6.9)$$

in which, v_s is the sedimentation velocity (m s^{-1}), g is the gravitational acceleration (m s^{-2}), $\Delta \rho$ the density difference between the phases (kg m^{-3}), d_{drop} the droplet diameter (m), and η_c the viscosity of the continuous phase (Pa s or kg m^{-1} s^{-1}). From this equation, it directly follows that a zero density difference makes the emulsion stable with respect to creaming; however, this is mostly not a practical solution in emulsion preparation since certain components are needed for appreciable rheological and other sensorial properties. Further, it is eminently clear that using small droplets will largely contribute to emulsion stability, that aggregates strongly enhance instability by creaming and that to a lesser extent viscosity of the continuous phase influences sedimentation. This also explains why the emulsions that are currently in the market have very small droplets, and mostly have a very viscous continuous phase.

When a droplet is not free to sediment due to the presence of other droplets (that can also move freely) the overall sedimentation velocity becomes lower. In literature, various authors have published equations that cover this behaviour for a range of dispersed phase fractions (for a good summary see Walstra 2003, 1996). Here we use the so-called Krieger-Dougherty equation (Equation 6.10). With the following equation, the ratio between the Stokes velocity of a single droplet and the velocity of droplets in the presence of others can be estimated.

$$\frac{v}{v_s} \approx \left(1 - \frac{\varphi}{\varphi_{max}}\right)^{k \cdot \varphi_{max}} \tag{6.10}$$

In which v is the velocity in a swarm of droplets (m s^{-1}), v_s the Stokes velocity for a single droplet (m s^{-1}), φ the volume fraction of dispersed phase (-), φ_{max} the maximum volume fraction of dispersed phase (-), and k a proportionality constant. For a k-value of 6.5, this equation has been validated for volume fractions of up to 40%, and it is a good intermediate choice compared to other models (Walstra, 1996).

The presence of droplets influences other droplets in their creaming/sedimentation behaviour, and this effect becomes very clear at higher volume fractions of dispersed phase (i.e. 0.3-0.4, Figure 6.6). Beware that the approach shown here is a starting point for estimating appropriate conditions for emulsion stability; it will always need to be followed up by experiments.

In literature, various sources use centrifuges to speed-up sedimentation, but you have to be careful in translating these data to a situation under gravity conditions. In the latter case, Brownian motion is of importance, while this is much less the case in a centrifuge. Translating one situation to another is notoriously difficult, although it is completely understandable that accelerated shelf-life experimentation is necessary to keep the development time for new products within limits.

Another aspect that is described to have influence on emulsion stability is Ostwald ripening, which occurs if there are droplets of different sizes (Laplace pressure difference) and the dispersed phase has some solubility in the continuous phase and can pass the interfacial layer. In that case, this component will move from an area

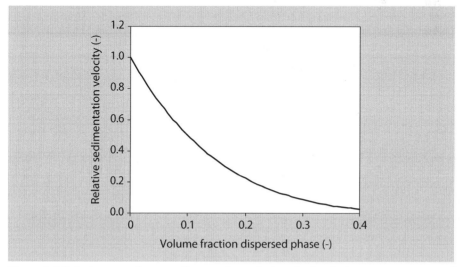

Figure 6.6. Relative creaming or sedimentation velocity of a droplet in a swarm compared to a single droplet as function of the volume fraction of droplets.

with high Laplace pressure (small droplet) to low Laplace pressure, leading to growth of the largest droplets and emulsion coarsening. This effect is very relevant for foams, but it has also been mentioned for emulsions (Walstra, 1996), although it is not expected to be the major cause of emulsion stability.

One last note on emulsion stability; for droplet coalescence, the film between droplets needs to break. Although various groups are investigating this process, it is still far from understood, therefore, we will not discuss this here. Some thoughts on dynamic interfacial tension (which implies surfaces that are not completely covered) are presented in the outlook section.

6.3. Emulsion preparation

6.3.1. Equipment

Emulsification in industry mostly takes place by applying a lot of brute force to the pre-emulsion. There are many different machines available, and reviewed in literature by e.g. Walstra (2003), Behrend and Schubert (2001), Schubert and Armbruster (1992), Arbuckle (1986) and Brennan (1986), Karbstein and Schubert (1995). For more detailed information, we would like to refer to these papers; here we focus on the effect of the created flow patterns on droplet formation.

Three methods that are used in industry are depicted in Figure 6.7. In the high pressure homogenizer, depicted on the left, the pre-emulsion is pressurized and pushed through a tiny hole; the local shear induces droplet breakup. The rotor-stator system consists of two concave elements, of which one rotates, and that create the necessary shear for droplet break-up. In the ultrasound equipment, the ultrasound creates cavitation bubbles that collapse, and through this action, droplets

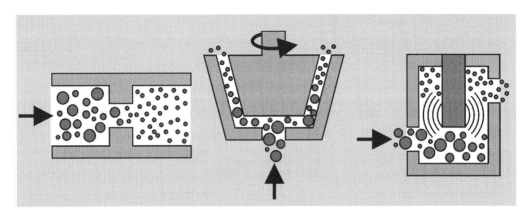

Figure 6.7. Schematic representation of classic emulsification methods, from left to right, high pressure homogenizer, rotor-stator system, ultrasound (Van der Zwan, 2008).

are submitted to shear. In general, when a liquid flows around a droplet, it induces a shear force onto the droplet, which will break if the exerted force is sufficiently large. Therefore, most methods are designed to generate a flow field that is very strong and that acts on a very small volume, through which the emulsion passes.

Rotor-stator systems

Various designs of rotor-stator systems can be found in practice, ranging from stirred vessels to colloid mills. The simplest form is the stirred tank that exerts rather erratic shear to the droplets and results in very polydisperse coarse emulsions with droplets > 10 μm. The energy density of the treatment is low; therefore, small droplets are outside the reach of this equipment. In a colloid mill, the central cone is static, while the outside cone rotates at high speed. This system can be operated in continuous mode with the pre-emulsion entering from the bottom and the fine emulsion leaving from the top. The distance between both parts is chosen in such a way that the distance is small, and shearing optimal. A variation to the colloid mill is the so-called toothed mill; it differs in that the rotor and stator both have openings, and because of this, the flow is turbulent.

The flow field in between the two elements of a colloid mill becomes very intense when the distance between them is very small; sometimes much less than a millimetre. Due to the high rotation rate, and the small size of the gap, the shear forces are very intense, and small droplet sizes can be realized. It depends on the viscosity of the mixture at what regime the mill operates. When the mixture is highly viscous, the flow will be laminar. The transition towards turbulent flow is given by the Reynolds number, which characterizes the flow:

$$\mathrm{Re} = \frac{\rho\, vL}{\eta}; \quad \mathrm{Re}_{cr} \approx 370 \tag{6.11}$$

Here, L is the gap width between rotor and stator, and v is the tangential speed of the rotor (m/s); the density ρ and viscosity η are the values for the mixture. When $\mathrm{Re} < 370$, the flow will be laminar. When it is larger, the flow will start to become turbulent.

High-pressure homogenizers

During passage of the emulsion through the very small hole, the liquid is subjected to extensional flow, turbulence, and possibly cavitation (imploding bubbles: see ultrasound), which all result in droplet break-up. Very intense fields can be reached, by using pressure differences of 10-50 MPa. Small, lab-scale homogenizers will operate in the laminar flow regime; industrial-scale systems will operate in the turbulent regime. The transition between these two regimes is once more given by the Reynolds number (now, L is the gap width, and v is the average liquid velocity in the gap):

$$\text{Re} = \frac{\rho v L}{\eta}; \quad \text{Re}_{cr} \approx 1500 - 3000 \tag{6.12}$$

In industry, two designs prevail:
- *Valve systems*: these are used especially in the dairy industry. The narrow gap is created here with a valve, which is pressed shut with a disk and a strong spring. Pressing the liquid through the valve lifts the disk somewhat: the force of the spring ensures that the gap width will remain very small. For a lab-scale system, the gap width will be of the order of 1 μm; for industrial-scale systems, it will be 10-40 μm (Smulders *et al.*, 2000).
- *Nozzle systems*. other systems do not have a valve, but simply have a small opening through which the liquid is squeezed; in some systems, two or more openings are placed in series. Sometimes the liquid is split up and then recombined through small holes, giving similar action.

Ultrasound

Ultrasound (i.e. sound with frequencies higher than 20 kHz) is generated by an actuator (comparable to a small loudspeaker), which vibrates with the specified frequency, resulting in a pattern of fluctuations (standing waves or pressure propagation). When the sound is sufficiently intense, the pressure fluctuations will become so large, that in small regions, the pressure becomes lower than the vapour pressure of water. This will induce the formation of small bubbles, which will implode almost immediately again. This implosion will cause very intense, local turbulence, which can break up the droplets.

The mixture is led around the actuator, into the volume that has the strongest field, and is then led away. The intensity of the ultrasound quickly declines away from the actuator; therefore, the treatment chamber has to be relatively small. This makes the technology not suited for very large-scale production; it is used commercially for low-volume products. For products that allow its application (unsaturated fats are notoriously unstable in ultrasound), use of ultrasound is an excellent way of generating very fine emulsions, due to the high intensity of the ultrasonic field that can be generated.

6.3.2. Flow in emulsification equipment

Depending on the flow inside the emulsification equipment, different mechanisms for droplet break-up may occur; ranging from rotational, and elongational, to turbulent flow. Whether a droplet is broken depends on the ratio between the external stress exerted by the continuous phase and the internal stress; the interfacial tension will resist deformation. This ratio is called the Weber number, and in general, it is defined as:

$$We = \frac{External,\ disruptive\ stress}{Internal,\ coherent\ stress} \tag{6.13}$$

For situations where the droplets are subjected to the disruptive stress for a very short time (e.g. in turbulent flow), the viscosity of the internal phase will cause the droplet to react slowly to the external stress, and the definition has to be extended:

$$We = \frac{External,\ disruptive\ stress}{Internal,\ coherent\ stress + coherent\ viscous\ forces} \qquad (6.14)$$

Laminar flow

If flow is applied, the droplet will feel the flow around it, and this will deliver the external, disruptive force for break-up. For laminar flow, two cases can be distinguished, simple shear flow and extensional flow. Both types are schematically depicted in Figure 6.8.

Simple shear flow
One encounters simple shear flow during flow through a tube, or flow over a planar surface. A droplet of size R_d subjected to simple shear flow will be distorted due to the stress exerted on the droplet. The internal, coherent stress can be estimated with the help of the Laplace pressure in the droplets $2\sigma/R_d$, which is the pressure that the interface exerts and keeps the droplet together. The disruptive stress can be estimated through:

$$\tau_{ext} = \eta_c \left(\frac{dv}{dz}\right) = \eta_c(\dot{\gamma}) \qquad (6.15)$$

Here, $\dot{\gamma}$ is the shear rate that is applied (s^{-1}), η_c is the viscosity of continuous phase (Pa s), v is the velocity of the continuous phase (m s^{-1}) and z is a position (m); and in combination with the Laplace pressure (Equation 6.4), this yields a relation for *We*:

$$We = \frac{\eta_c \dot{\gamma} R_d}{2\sigma} \qquad (6.16)$$

In order to achieve break-up of a droplet, this *We*-number should exceed a certain critical value, called the critical *We*-number:

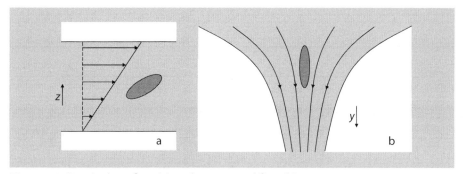

Figure 6.8. Simple shear flow (a) and extensional flow (b).

$$\frac{\eta_c \, \dot{\gamma} \, R_d}{2\sigma} > We_{cr} \qquad\qquad (6.17)$$

It is found that the exact value of We_{cr} is not completely constant, but is dependent on the ratio of the viscosities of the dispersed and the continuous phase, as indicated in Figure 6.9. This is because the droplet will deform more when the dispersed phase viscosity is lower, which will give a higher Laplace pressure and a lower external stress.

Extensional flow
This type of flow takes place when the liquid is squeezed into a small opening. It occurs when the liquid enters (or exits) a channel, or is pushed through a small hole (e.g. high-pressure homogenization). A droplet in extensional flow will also experience a drag force exerted by the flow; only now the external force exerted on the droplet is not equal to $\eta_c(dv/dz)$ but equal to $\eta_c(dv/dy)$, where y is the coordinate in the direction of the extension. We can use the same relations as for simple shear flow, only in this case the value of We_{cr} is different. So, also here:

$$\frac{\eta_c \, \dot{\gamma} \, R_d}{2\sigma} > We_{cr} \qquad\qquad (6.18)$$

In general, extensional flow is more effective than simple shear flow in disrupting a droplet (in the latter case, energy is used to rotate the droplet). Therefore, the We_{cr}-number for extensional flow is smaller than for simple shear flow. Figure 6.7 gives an impression of the values both for simple shear flow and for extensional

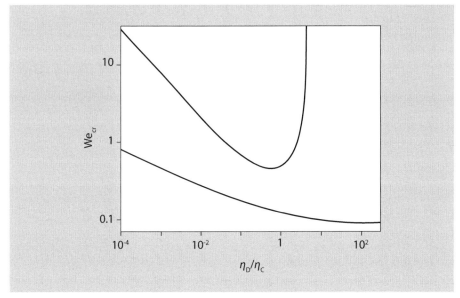

Figure 6.9. Critical We-numbers for laminar flow conditions: plain shear flow (upper curve), and for extensional flow (lower curve) (Walstra, 2003).

flow. In practice, one will always have a mixture of the two types of flow; and the critical *We*-number will have an intermediate value. Still the diagram is very useful to calculate best- and worst-case scenarios.

Turbulent flow

For example at the high pressures applied during high pressure homogenization, the flow will no longer be laminar, and turbulence will occur; which implies that the liquid will start moving in a chaotic way and swirls and eddies will occur. The transition from laminar to turbulent flow is characterized by the Reynolds number; the critical Reynolds number depends on the geometry and the product properties (see Equations 6.11 and 6.12). Turbulent flow can be very effective in breaking up droplets into smaller ones. The swirls and eddies are typically smaller than the droplets, and they deform and break-up the droplets (see Figure 6.10).

Given the chaotic nature of the process, it is more difficult to find a critical Weber-number, since the exact local flow-conditions cannot be determined. What is usually done, is that the power density is taken as a measure for the intensity of the swirls and eddies. The power density (symbol ε, with unit W/m^3) is the amount of energy per m^3 that is used to establish the flow pattern. The larger the power input, the more intense and smaller the swirls and eddies will be.

When the turbulence is not too high, the main force on the droplets is caused by the shear imposed by the surrounding eddies. With the help of the Kolmogorov theory for turbulent flow, the external, disrupting force can be estimated as $\tau_{ext} - \sqrt{(\varepsilon \cdot \eta_c)}$, in which ε is the power density (Pa s^{-1}), η_c the viscosity of the continuous phase (Pa s), and R_d the droplet diameter. The critical Weber number now becomes:

$$We_{cr} = \frac{\tau_{cr} R_d}{2\sigma} \approx \frac{\sqrt{\varepsilon \eta_c} R_d}{\sigma} \qquad (6.19)$$

Figure 6.10. Break-up by turbulent flow: the swirls and eddies in the liquid impinge on the droplet, deforming it, which leads to break-up into smaller droplets.

As a first approach, one can assume that We_{cr} should be around one, and the droplet size can be estimated.

When the turbulence becomes very intense, the shear forces exerted by the swirls and eddies are not the dominant force anymore; rather the inertia of the liquid impinging on the droplets become the disruptive force. With the help of the Bernouilli equation, we can estimate the external, disruptive force as:

$$\tau_{ext} \approx \varepsilon^{2/3} R_d^{5/3} \rho_c^{1/3} \tag{6.20}$$

With ρ the density (kg m^{-3}). The equation for the critical Weber number then becomes:

$$\tau_{ext} \approx \varepsilon^{2/3} R_d^{5/3} \rho_c^{1/3} \tag{6.21}$$

Again, by assuming that We_{cr} will be around unity, one can have an estimate of the droplet size obtained.

The transition from viscous-dominated break-up to inertia-dominated break-up takes place when the droplets are larger than (Walstra, 2003):

$$R_d > \frac{\eta_c^2}{\sigma \rho_c} \tag{6.22}$$

When relating this back to the previously described emulsification techniques, flow in a stirred tank is mostly turbulent, but not very intense; therefore, it will operate in the viscous-force dominated regime. Flow in a colloid mill can be laminar or turbulent, depending on the viscosity of the product. For water-in-oil emulsions, η_c is large and break-up will usually be viscous force dominated; for oil-in-water emulsions, it depends on the droplet size. Flow in a toothed mill is often turbulent, but again the break-up mechanism depends on droplet size and viscosity of the continuous phase. Large-scale high-pressure homogenizers feature turbulent flow and inertial forces dominate, while laboratory scale homogenizers operate in the laminar regime. Therefore, translation of results obtained on laboratory scale is not straight forward (Smulders, 2000; Walstra and Smulders, 1998). Emulsification with ultrasound is always based on inertial forces, created by the cavitations of the vapour bubbles.

6.4. Emerging emulsification technologies

At the moment, emulsification research is a lively research area with many different new emulsification technologies being proposed, all with their own pros and cons. Before diving into the latest additions to the field, we first will discuss membrane emulsification, which has been around for 2 decades, and compare its performance with some of the more classic technologies regarding energy usage. Next, we will present a selection of microfluidic techniques that have been shown to give very monodisperse emulsions, and give design guidelines for these processes. In the last section, all methods will be compared, and an outlook on emulsification technology will be given.

6.4.1. Membrane emulsification

A membrane is a porous structure that is mostly used for separation but can also be used to make emulsions and related products. Two of the most frequently used membranes for emulsification are the so-called SPG (Shirazu Porous Glass; Nakashima and Shimizu, 1986) and ceramic membranes (Schröder *et al.*, 1998; Schröder and Schubert, 1999; Vladisavljevic and Schubert, 2003). Both membranes are depicted in Figure 6.11; the SPG membrane consists of a matrix of interconnected pores that have similar size all through the membrane, while the ceramic membrane consists of a carrier with large pores onto which a layer with small pores is deposited.

The membranes are mostly applied either in cross-flow mode (Nakashima and Shimizu, 1986) or in pre-mix mode. During cross-flow emulsification (see Figure 6.12a), the to be dispersed phase is pressed through the membrane where it forms small droplets on top of the membrane, that are consecutively sheared off by the cross-flowing continuous phase once they have grown to a certain size as will be explained later. This approach can be used for oil-in-water, and water-in-oil emulsions, and more complex systems.

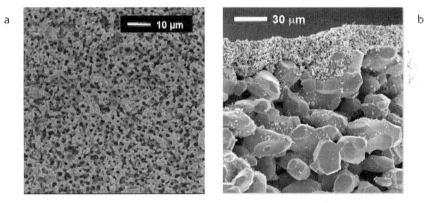

Figure 6.11. (a) Shirazu Porous Glass (SPG) membrane with interconnected tortuous pores of similar size all through the membrane. (b) Ceramic membrane with an open support structure and much finer top-layer.

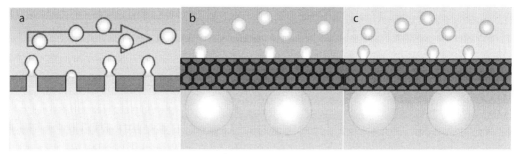

Figure 6.12. Schematic representations of (a) Cross-flow membrane emulsification, in which the cross-flowing continuous phase shears-off the droplets that are formed through pressurization of the to-be-dispersed phase. (b) Pre-mix emulsification, in which a coarse emulsion is broken up into smaller ones by passage through a membrane. (c) Pre-mix emulsification through phase inversion.

During pre-mix emulsification, the big droplets of a pre-emulsion are broken up into smaller ones, as is the case in the classic emulsification techniques (see also Figure 6.12b). The membrane needs to be chosen in such a way that it is wetted by the continuous phase of the emulsion (so hydrophilic membrane for oil in water emulsions, and hydrophobic membranes for water in oil emulsions). In some cases, it is possible to use phase inversion during pre-mix emulsification. In that case, the membrane should be compatible with the to-be-dispersed phase. If you start with an oil-in-water pre-mix, the membrane needs to be hydrophobic; the oil droplets will wet the membrane, and the continuous water phase will be converted into the dispersed phase during passage through the membrane (Suzuki *et al.*, 1999). If this mode of operation is possible, emulsions with very high dispersed phase fraction may be obtained, but this strongly depends on the components in the emulsion mix, and on their interaction with the membrane.

To generate the required shear for droplet detachment, alternative designs have been proposed in literature. To name a few, Stillwell and co-workers (2007), investigated a stirred cell in which a membrane was mounted at the bottom of a vessel and a stirrer was used to put the continuous phase in motion. The resulting emulsions were rather polydisperse due to the differences in shear across the membrane. Eisner (2007), Eisner-Schadler and Windhab (2006), Aryantia and co-workers (2006), and Yuan and co-workers (2008) used a different approach and rotated (metal) membranes to shear off the droplets. This results in better control over droplet size. However, when compared to regular cross-flow emulsification, the droplets are much larger due to the technical limitations of the metal membranes that need to be used in such a device. We will not discuss these methods further here.

Cross-flow emulsification has been described abundantly for some time (e.g. by the inventors in the group of Nakashima from Japan, 1991). Good reviews that show the options for cross-flow membrane emulsification are available and we recommend the work of Joscelyne and Trägårdh (2000, general review), Charcosset and co-workers

(2004, general review), Vladisavljevic and Williams (2005, general overview with many products), Van der Graaf and co-workers (2005a, double emulsions), and Charcosset (2009, specific for food). The technology did not remain at an academic level; Morinaga Company experimented with membrane emulsification for food production (Katoh *et al.*, 1996), and launched a product (viz. a low fat spread) based on it. Most information on membrane emulsification is available for oil-in-water emulsions, but also some authors have shown work on water-in-oil emulsions, e.g. Vladisavljevic *et al.* (2002). On the other hand, pre-mix emulsification is not well documented; only very recently, a review became available by Nazir and co-workers, 2010.

Cross-flow membrane emulsification

The process of cross-flow membrane emulsification has been investigated in detail, and can be brought back to a force balance that acts on the forming droplet. As was the case for droplet break-up in Section 6.3, the interfacial tension and drag force determine when a droplet will be formed (Peng and Williams, 1998). The cross-flowing continuous phase exerts a drag force on the droplet, which can be estimated with the help of Stokes' law:

$$F_d = -1.7\left(6\pi\eta_c R_d v_c\right) \tag{6.23}$$

in which v_c is the velocity of the continuous liquid at half-height of the droplet (m s^{-1}), η_c is the viscosity of the continuous phase (Pa s), and R_d is the radius of the still just connected droplet (m). The factor 1.7 stems from the non-homogeneous velocity field around the droplet (see Figure 6.13).

The local velocity, v_c, can be estimated from the shear stress at the wall, τ_w, since:

$$\tau_w = -\eta_c \frac{dv_c}{dz} \approx -\eta_c \frac{v_c}{R_d} \tag{6.24}$$

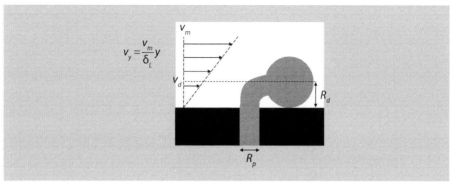

Figure 6.13. Highly schematized graph that shows forces acting on a deformed droplet in cross flow.

Substituting this in Equation 6.21, gives:

$$F_d = -10.2\pi R_d^2 \tau_w \tag{6.25}$$

The force that retains the droplet to the pore mouth is generated by the interfacial tension, σ (N m^{-1}), and the resulting force equals:

$$F_\sigma = 2\pi\sigma R_p \tag{6.26}$$

in which σ is the interfacial tension, R_p is the radius of the pore (m). The neck, which retains the droplet to the pore mouth, is roughly of the same thickness as the pore (and smaller than the droplet size). At the point where the droplet is still just attached to the pore, both forces are equal, and the ratio of R_d over R_p, can be calculated (as described in Equation 6.27).

$$\frac{R_d}{R_p} = \sqrt{\frac{\sigma}{5.1 \cdot \tau_w \cdot R_p}} = \sqrt{\frac{1}{5.1 \cdot Ca}} \tag{6.27}$$

Ca, the Capillary number, is defined as $\tau_w \cdot R_p / \sigma$ and it is the ratio between the interfacial tension force and drag force. It is similar to the *We*-number used earlier. As soon as the drag force exceeds the interfacial tension force, the droplet will be detached and taken up by the continuous phase. This equation can be very useful, since it links the process conditions to the droplet size that will be obtained; small membrane pores also lead to small droplets if the cross-flow velocity is chosen appropriately.

However, be aware that the equation is only to be used if the two described forces are the only ones that are acting. For SGP membranes, it is expected that spontaneous inflow of the continuous phase in the membrane takes place, therewith facilitating droplet snap-off similar to microchannel emulsification as described in the respective section. Further, the interfacial tension that is needed in these equations may be hard to estimate. Mostly one uses the static interfacial tension of two liquids with the surfactant under equilibrium conditions, but this may greatly underestimate the actual value (and therewith also the droplet size). If the surfactant is not fast enough to cover the interface, the droplets may become considerably larger.

Pre-mix emulsification

As mentioned, there is not a lot known about pre-mix emulsification, and this has to do with the very dynamic processes occurring inside the membrane. Some were visualized by Van der Zwan and co-workers in microfluidic devices (2006), and amongst others: snap-off due to localized shear, break-up due to interfacial tension effects, and break-up due to steric hindrance were found to occur. Besides, it was noted that interaction between droplets is taking place resulting in break-up, but unfortunately, this is not understood yet. It is expected that through microfluidic studies this terrain will become much better charted (Link *et al.*, 2004).

Comparison of membrane emulsification with 'classic' emulsification techniques

When comparing membrane emulsification with the more classic techniques that were described in the previous section as is done in Figure 6.14 (Nazir *et al.*, 2010), it is clear that cross-flow emulsification is a much milder, less energy consuming technique than the high pressure homogenizers reported by Lambrich and Schubert (2005). In some cases, the energy density is orders of magnitudes lower. The energy density of cross-flow membrane emulsification is dependent on the number of droplets that need to be produced. Each volume fraction has its own corresponding line, unlike other emulsification techniques, in which pressurization of the whole volume determines the amount of energy needed. Pre-mix emulsification seems to be more in line with the traditional techniques, but may be useful in the production of relatively large droplets. Also, this technology is not as developed as cross-flow emulsification, and improvements are expected in the near future.

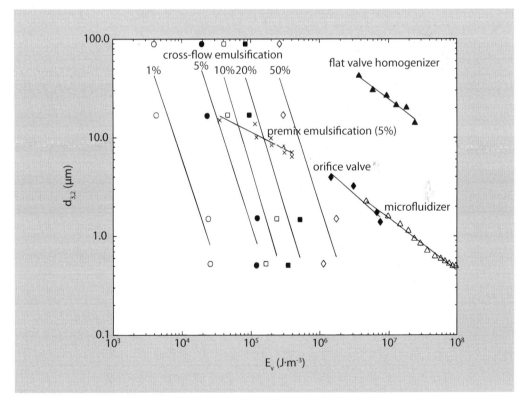

Figure 6.14. Energy efficiencies of various emulsifying processes. Cross-flow emulsification (Lambrich and Schubert, 2005), (○) 1%, (●) 5%, (□) 10%, (■) 20% and (◇) 50%; (×) pre-mix emulsification (5%) (Van der Zwan et al., 2008); high pressure homogenization (Lambrich and Schubert, 2005), (◆) orifice valve, (▲) flat valve homogenizer and (△) Microfluidizer (all 30%). (Reprinted with permission from Elsevier from Nazir et al., 2010).

Karin Schroën and Remko Boom

6.4.2. Emulsification with microfluidic devices

As mentioned previously, the emulsification field is developing very fast at the moment, and this implies that many different microfluidic designs are used to make emulsions. It would lead much too far to even try to review them all. We limit ourselves to T- (and Y) shaped junctions in which the shear is used to make droplets, and microchannel and EDGE (Edge-based Droplet GEneration) emulsification in which the Laplace pressure differences lead to droplet formation. The T- and Y-shaped junctions have similar mode of action as cross-flow membrane emulsification. In the next section, we will discuss upscaling of various microfluidic devices and membrane emulsification, in the light of large-scale production, and discuss related products that could be made with these technologies.

Shear-based droplet formation systems

During cross-flow emulsification, it is hard to observe droplet formation, if not even completely impossible. Therefore, various efforts were made to idealize a membrane or even a pore and learn from that. Pioneering work was done with so-called microsieves invented by Van Rijn *et al.* (1995; Van Rijn, 2004), which are extremely flat, porous plates produced with photolithographic techniques, which have uniform pore size; see Figures 6.15 and 6.20 (microsieves will be discussed in detail in the section on up-scaling). These devices can be placed underneath a microscope and droplet formation can be observed from the top (Abrahamse *et al.*, 2002; Van der Graaf *et al.*, 2004; see Figure 6.15a).

So-called T-junctions (and later Y-junctions; Steegmans *et al.*, 2009a) were used to investigate droplet snap-off from the side (Van der Graaf *et al.*, 2005b; Figure 6.15b),

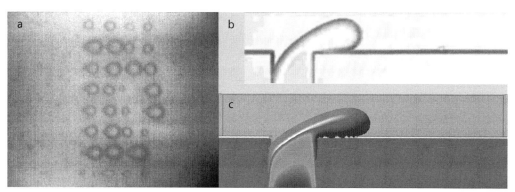

Figure 6.15. (a) Top view of a microsieve with droplets forming from the pores (Van der Graaf et al., 2004). (b) Side view image of a T-shaped junction with the oil pressurized from the bottom channel into the horizontal channel with the cross-flowing continuous phase (Van der Graaf and co-workers, 2005b). (c) Result from Lattice Boltzmann simulation for conditions as in the image above. (Van der Graaf et al., 2006; reprinted with permission from ACS)

and compared to computational results (Van der Graaf *et al.*, 2006; Figure 6.15c). In a T-junction, the to-be-dispersed phase is pressurized into a channel with the cross-flowing continuous phase, where it slowly protrudes into the continuous phase channel. Due to the shear of the continuous phase, the to-be-dispersed phase will be carried further into the continuous phase channel, and ultimately the droplet will snap off. A similar force balance holds as was described for cross-flow membrane emulsification; the shear force and the interfacial tension force work opposite, and at some stage, the shear force becomes thus large that it exceeds the interfacial tension force, and the droplet breaks off.

For Y-shaped junctions, the force balance is the same as described earlier, and the Capillary number can be used to predict the size of the droplets formed (Steegmans, 2009a). However, for T-shaped junctions, the situation is more complex. For this design, two stages can be distinguished during droplet formation (which is also reported for bubbles in T-junctions; Van Steijn, 2010). Droplet formation takes place in two stages, one in which the droplet obtains a certain size (which is determined by the Ca-number i.e. the force balance) followed by a stage during which the droplet is still fed by the neck that keeps it connected to the feed channel. Basically, it takes time for the neck to break and during this time the droplet will grow bigger. For Y-junctions, Steegmans *et al.* (2009a) found:

$$\frac{r_{drop}}{z} = 0.167 \left(\frac{\frac{w}{z} - 0.89}{\frac{w}{z} + 0.79} \right)^{0.5} \cdot Ca^{-0.5} \qquad \text{Y-junctions} \qquad (6.28)$$

In which, r_{drop} is the droplet radius (m), w is the width of the channel (m), z is the depth of the channel (m), and Ca is defined as $\eta_c v_c / \sigma$, with η_c the viscosity of the continuous phase (Pa s), v_c the average velocity of the continuous phase (m s^{-1}), and σ the interfacial tension (N m^{-1}). For Y-junctions, the size of the droplet is only determined by the conditions of the continuous phase (and obviously the design of the microfluidic device). For T-shaped junctions the situation is different (Steegmans *et al.*, 2009b):

$$\frac{r_{drop}}{z} = \frac{\sqrt{r_{growth}^3 + \frac{3}{4\pi} t_{neck} \cdot \Phi_d}}{z} \qquad \text{T-junctions} \qquad (6.29)$$

In which, r_{growth} is the radius to which the droplet needs to grow to satisfy the force balance (m), t_{neck} is the time needed for the neck to break (s), and Φ_d the disperse phase flow (m^3 s^{-1}). Contrary to the Y-junctions, the droplet size now has become a function of both the continuous and the to-be-dispersed phase. At very low dispersed phase flow rate, the influence of the 'necking' process will become negligible, and the basic force balance will determine the size of the droplets.

The studies on shear-based emulsification have yielded valuable insights in emulsification processes; both through experimentation and simulation (See also

Figure 6.15b and c for a comparison between experimental results and Lattice Bolzmann simulations (Van der Graaf *et al.*, 2006)). The scaling relations that are currently available are of great use for (re)designing emulsification processes, although in scaled up versions other aspects also play a role as will be explained in the respective section.

Spontaneous droplet formation systems

Microchannels

The microchannel was first introduced by Kikuchi and co-workers in 1992 for the observation of blood cells, and was later, in 1997, adjusted for emulsion formation by Kawakatsu and co-workers (2007), within the group of Nakajima in Japan. Here, the technology was further investigated, which resulted in many papers during the years, from e.g. Sugiura *et al.* (2001, 2002, 2004) and Kobayashi (some are given in the list of references).

The microchannel consists of a channel through which the to-be-dispersed phase is introduced onto a wider shallow area called the terrace. The to-be-dispersed phase will invade the terrace and form a disk shape; eventually it will reach the end of the terrace where the liquid can flow into a deeper channel called the well. The entire structure is covered with a flat plate; an artist's impression of the process is shown in Figure 6.16.

The driving force behind this process is a Laplace pressure difference between the pancake structure on the terrace (one principle curvature), and the connected

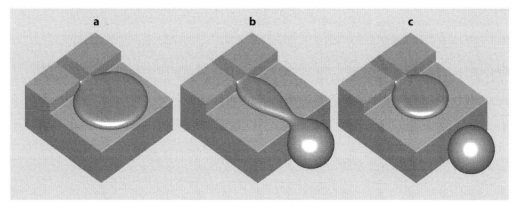

Figure 6.16. Spontaneous droplet formation in microchannel systems. In (a), the to-be-dispersed phase is pushed through the channel on the left (in reality this channel is very long) onto the terrace where it forms a disk. As soon as the disk reaches the end of the terrace (b), the liquid can flow in a deeper part and assume a spherical shape. As soon as the size of the developing droplet exceeds a certain size, a droplet will be formed (c). (Van der Zwan et al., *2009a; reprinted with permission from ACS).*

spherical droplet (two identical curvatures), which has an energetically favourable shape. The process can be described based on the two pressure differences in the system, one between the applied pressure and the pressure in the neck that keeps that droplet connected, and one between the neck and the droplet. The oil flow that is a result of these pressure differences can be formulated as follows (Van Dijke *et al.*, 2008):

$$\Phi_{ch,n} = \frac{\pi R_c^4 \left(P_{app} - \dfrac{\sigma \cos\theta}{R_c} \right)}{8\eta L_{ch}} \qquad (6.30)$$

The flow from channel to the neck ($\Phi_{ch,n}$ m^3 s^{-1}) is a Poisseuile flow through a channel of length L_{ch} (m), and for a liquid with viscosity η (Pa s), which is governed by the difference in applied pressure (P_{app}: Pa) and the Laplace pressure in the neck. This latter pressure is determined by the interfacial tension σ, (N m^{-1}), the contact angle θ and the height of the channel R_c (m), which determines the principle curvature of the neck (the other curvature is infinite).

For the flow from neck to droplet ($\Phi_{n,d}$ m^3 s^{-1}), the equation is similar, but now the pressure differences are the Laplace pressure in the neck and in the droplet (which is determined by the interfacial tension and the droplet size (R_d: m), and the flow takes place through a neck of length L_n (m).

$$\Phi_{n,d} = \frac{\pi R_c^4 \left(\dfrac{\sigma \cos\theta}{R_c} - \dfrac{2\sigma}{R_d} \right)}{8\eta L_n} \qquad (6.31)$$

The length L_n over which the pressure difference acts needs to be defined, and how this can be done is described in Van Dijke *et al.* (2008). Here, we only want to use the equations to illustrate the typical behaviour of microchannels; an example is shown in Figure 6.17.

At low to-be-dispersed-phase velocity (i.e. applied pressure), the droplets will get a certain size, and will snap off because the flow through the channel becomes lower than the flow through the neck at some size of the droplet; which implies the supply is not large enough to keep up with the demand from the droplet. If the pressure is even higher, the droplet size at which droplet break off will occur will become slightly higher, but at really high pressure, the flow through the channel will no longer be able to become lower than the flow through the neck. The droplets will stay connected and the size will increase sharply (this is the so-called blow-up point, as can also be seen from the abrupt change in slope of the lines in Figure 6.17).

Equations 6.30 and 6.31 hold, as long as the continuous phase can flow fast and freely onto and from the terrace, and the pressure differences used in these equations are only determined by the to-be-dispersed phase. This will become different if the continuous phase is more viscous; at some viscosity, the flow of the continuous phase will impose an external pressure difference, and because of this, the droplet size will

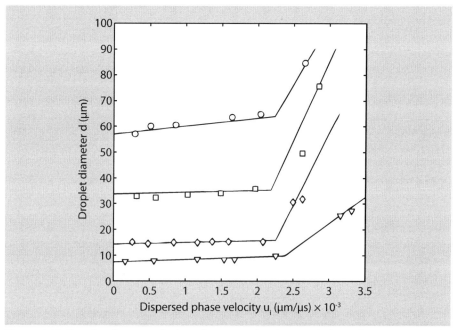

Figure 6.17. Droplet diameter d *plotted against the dispersed phase flow velocity* u_i *for 4 different microchannel geometries; the lines in the graphs are strictly to guide the eye. (Data from Sugiura* et al., *2002 as used in Van der Zwan, 2009a; reprinted with permission from ACS).*

become larger. These aspects have been investigated experimentally (as illustrated in Figure 6.18). Below a certain viscosity ratio between continuous and to-be-dispersed phase the droplet size increases (keep in mind that this is not a blow-up effect, this is just an increase in droplet size in the stable droplet formation regime). This effect was also discussed by Vladisavljevic *et al.* (2010) regarding straight-through microchannels.

The fact that so much is known about microchannels is also due to various computational analyses that have helped to understand the observed effects. The approaches that have been published are diverse. A geometrical analysis was carried out by Van der Zwan *et al.* (2009a), CFD (computational fluid dynamics) type simulations were done by Kobayashi *et al.* (2004, 2006); for straight-through systems, and for microchannels Van Dijke *et al.* 2008 used Star-CD, also to include the effect of viscosity ratio. Van der Zwan *et al.* (2009b) used Lattice Boltzmann simulations for microchannels, and all these efforts have in their own right added to a better understanding of the droplet formation mechanism.

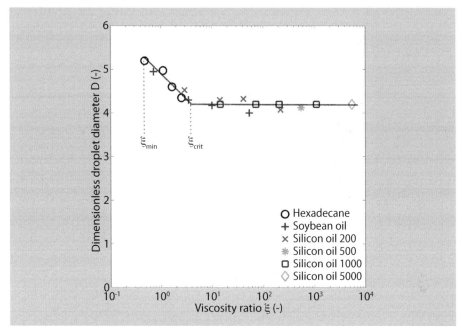

Figure 6.18. Dimensionless droplet diameter (droplet diameter over depth of the terrace, R_c), as function of the viscosity ratio between the continuous phase and the to-be-dispersed phase. (Van Dijke et al., 2010a; reprinted with permission from Springer Verlag).

EDGE systems

The acronym EDGE means Edge-based Droplet GEneration systems and the name is inspired by the pivotal role of the edge of a plateau in droplet formation. At first glance, this system may look similar to the previously described microchannel systems, but it has an essential difference. In a microchannel, the volume present on the terrace empties into one droplet. For EDGE devices, this is not the case; multiple droplets are formed simultaneously. A screenshot of a high-speed camera movie is shown in Figure 6.19.

The oil channel feeds the to-be-dispersed phase onto the shallow plateau (which is much wider than a terrace), and the oil invades it from a central point and spreads all over its width. When the oil-water interface reaches the edge of the plateau, where a deeper channel is present, local necks are formed, and droplets are formed. The mechanism behind this is a result of the different local pressures in the system as was described for microchannels. This system also shows blow-up behaviour, but overall it seems to be more stable and shows a droplet size that is independent of the applied pressure for a relatively wide range of pressures. In the stable droplet formation range, the size of the droplets only scales with the height of the plateau (H), or as an equation:

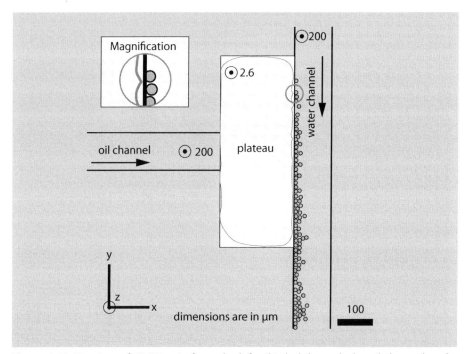

Figure 6.19. Topview of EDGE unit; from the left oil is led through the oil channel to the plateau and into the water channel after droplet formation at the edge of the plateau. A magnification of droplet formation is shown in the top left corner. Typical depths of the structure are indicated by the numbers with ⊙. (Van Dijke et al.*, 2010 a; reprinted with permission from Wiley).*

$$\frac{d_{drop}}{H} \sim 6 - 8 \tag{6.32}$$

Because of its simplicity, the EDGE design, the EDGE design is well suited for upscaling (see respective section): in principle, only a shallow area is needed that is connected to a deeper area. Further, it was noted that also here the viscosity ratio between the continuous and to-be-dispersed phase influences the droplet size (Van Dijke *et al.*, 2010b), but overall, the EDGE system is very robust also when food grade components are applied either for single or double emulsions, or foams (Van Dijke *et al.*, 2010c).

Scaling-out of microfluidic systems

Microfluidic devices are known for the monodisperse emulsions that they can produce; however, in order to become practically relevant, they need to be scaled-out by parallelizing many units. A membrane is in principle already an up-scaled version of a membrane pore, consisting of many pores in parallel, albeit not of uniform size.

Of most of the microfluidic devices presented in the previous section, out-scaled versions have been published, and ideas for design of out-scaled devices have been proposed, and experimentally tested on various scales. The most illustrative one is most probably the microsieve, which consists of a silicon plate that has been treated with photolithographic techniques, and contains pores of uniform size, and a chosen geometry. Three examples are shown in Figure 6.20.

In principle, microsieves can have a very high flux, because their resistance is low compared to that of other membranes; they can be as thin as 1 μm. Because of this, the required surface area for a specific membrane emulsification process suggested by Gijsbertsen-Abrahamse and co-workers (2004) is reduced from thousands of m^2 for a ceramic membrane, around a hundred m^2 for and SPG membrane, to around 5 m^2 for the microsieve. This brings membrane emulsification within practical reach: but please note that at the time of writing, microsieve technology is still in the development phase.

Some aspects need to be controlled very closely, such as the wettability of the microsieve that needs to be constant in time; if not the oil may stick to the surface and the emulsion process will be lost (Abrahamse *et al.*, 2001). The surface modification techniques developed by e.g. Arafat *et al.* (2004 and 2007), and Rosso *et al.* (2009), could help prevent these wettability changes. Further, Abrahamse *et al.* (2002) reported that the microsieves need to be relatively thick in order to prevent communication between the pores, which results in low pore activation. And besides, the pore should be positioned appropriately, in order to keep the forming droplets from interacting, and causing polydispersity. If all these aspects are covered, all pores can be activated, and monodisperse emulsions can be produced. Given the newness of the microsieve, it is expected that if it is to be used on an industrial scale, it will first be for a high quality (low volume) product.

The flow focusing devices that were first introduced by Anna and co-workers (2003) have been scaled up by Nisisako and Torii (2008; see Figure 6.21); the single structures are grouped around a central exit area, as depicted in Figure 6.21. All units contain two continuous channels on the outside, and one to-be-dispersed phase channel, positioned in the middle. Through the shear exerted by the continuous

Figure 6.20. Three microsieve designs with round and rectangular pores. (Images courtesy of Aquamarijn Microfiltration B.V.).

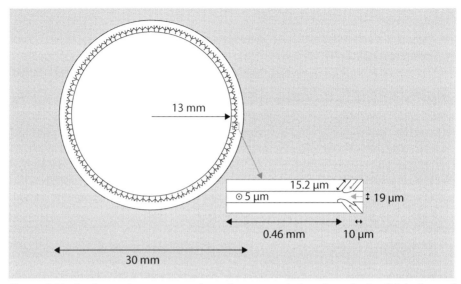

Figure 6.21. A schematic representation of an up-scaled co-flow device (Nisisako and Torii, 2008). All single units have two continuous phase channels that are positioned on the outside, and one to-be-dispersed channel positioned between these two channels.

phase, droplets are generated from the middle channel, and Nisisako and Torii have succeeded to operate around 150 units simultaneously. Please note that all units needed to be connected individually, so this was far from trivial to do, and this layout is expected to be difficult to scale up further. Obviously, the same approach could be used for T- or Y-junctions; however it makes more sense to arrange these channels differently, as was suggested by Steegmans *et al.* (2009; see also Figure 6.22). In that way, relatively compact microstructures can be obtained that also do not need the amount of tubing that is necessary for the Nisisako and Torii design.

Microfluidic devices that use spontaneous droplet formation have also been scaled up. The microchannels were taken a step further, by Kobayashi and co-workers,

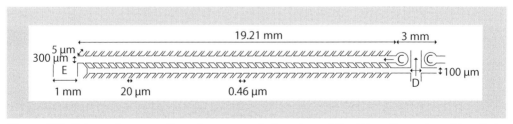

Figure 6.22. Schematic suggestion of an up-scaled version of the Y-junction; at position C the continuous phase enters and passes the Y-junctions where the disperse phase is sheared. (Steegmans et al., 2009).

Food product design

resulting in straight through emulsification devices (2002 and 2005a,b; see also Figure 6.23a). In general, the droplet size is determined by the dimensions of the terrace and the flow rate of the to-be-dispersed phase, not by the continuous phase and this makes this technology rather robust. Also for this technology, pore activation became an issue when narrow pores were used, and the design of the device needed to be adjusted.

Two other technologies need to be mentioned. First the technology of Nanomi that also uses silicon plates into which specific pores are made by photolithographic techniques (see Figure 6.23b for a screen shot during emulsification; Veldhuis, personal communication). Further, the previously mentioned EDGE technology has been scaled up to some extent (see Figure 6.23c), and tested (Van Dijke *et al.*, 2009). It was found that the entire chip produced droplets or bubbles of identical size. The chip could be operated for longer periods without any notable change in the emulsification process (Van Dijke *et al.* 2010c), and this makes EDGE technology an interesting option for scale-up.

6.5. Outlook and challenges for emulsification technologies

When comparing the classic emulsification methods with membrane/micro device emulsification as is done in Table 6.1, the conclusion needs to be that in spite of their high energy-usage, and the polydispersity of the produced emulsions, classic methods are still very much in use. This is because of the relatively low productivity of these new devices as illustrated in Table 6.2, and discussed later. In general, the classic methods are not sensitive to fouling, unlike methods based on membranes and micro technological devices, and this is a big pro for the classic techniques.

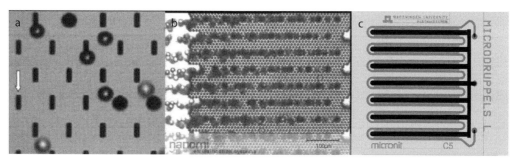

Figure 6.23. Up-scaled microfluidic devices that make use of spontaneous droplet formation. (a) Straight through microchannels (Kobayashi et al., *2005a,b), (b) Nanomi technology, (c) EDGE technology with 7 parallel long plateaus; the to-be-dispersed phase is introduced through the black circuit, the continuous phase meanders around the to-be-dispersed phase channels (dark grey line) between which the plateaus are situated (light gray).*

Table 6.1. Comparison of various emulsification methods.

Method	Productivity	Energy usage	Droplet size	Monodispersity	Fouling
High pressure homogenization	++++	++++	micron to submicron	+	0
Rotor-stator	+++	++++	microns	+	0
Ultrasound	+++	++++	micron to submicron	+	0
Premix membrane emulsification	++	++	micron to submicron	++	---
Cross-flow membrane emulsification	+	+	micron to submicron	+++	-
Micro technological emulsification	-	+	micron to submicron	++++	-

Number of + and - signs signifies the level of productivity, the level of energy usage, the level of monodispersity, and how resistant to fouling the emulsification method is. 0 indicates not affected by fouling.

Table 6.2. Estimation of the device volume V_{device} and required area A_{device} for oil dispersion of $1\,m^3{\times}h^{-1}$ for various shear-driven emulsification devices.

Emulsification device	V_{device} [L]	A_{device} [m²]
Y-junctions around circular area ID=1 mm and OD=3mm[1]	18-235	4-52
500 Y-junctions at one continuous-phase channel[1]	7-90	0.1-2
Microsieve[1]	0.05	0.1
SPG membrane[1]	2	0.3
Ceramic membrane[1]	30	7
EDGE[2]	875	35
Straight through microchannels (assuming all pores are active)		70-140

ID: internal diameter, OD: outer diameter of Nisisako and Torii design shown in Figure 6.21, (2008).
[1] Refers to Steegmans (2009: Figure 6.22).
[2] Refers to Van Dijke (2009: Figure 6.23c).

Membrane emulsification techniques have much lower energy consumption, and maybe even more importantly, they can render emulsions with considerably narrower droplet size distributions (Table 6.1). For as far as the actual droplet size is concerned, the membrane emulsification techniques are comparable with the classic technologies except for the rotor-stator systems which give larger droplets. The last entry in the table relates to micro technological devices, which are known for their extremely narrow droplet size distributions. Their energy consumption is expected to be comparable to cross-flow membrane emulsification or even lower, and therewith considerably lower than in the classic technologies. Detailed information on the emulsification methods can be found in the four cited works of the group of Professor Schubert: Karbstein (1995), Schröder (1999), Behrend (2001), Lambrich (2005), and in Gijsbertsen-Abrahamse (2003; Gijsbertsen-Abrahamse *et al.*, 2004), and in Eisner (2007).

Various issues need to be resolved for the membrane/micro device related methods. Given the current speed of the developments, including detailed understanding through visualization and modelling, and the versatility of the technology with respect to specialty products that cannot or can hardly be produced with classic technology, it is expected that emulsification with membranes and micro devices will find their place in the world of emulsification.

When comparing the various shear-based emulsification methods like cross-flow emulsification with spontaneous emulsification methods, it needs to be mentioned that the rate at which droplets are formed from one pore varies considerably. In shear-based emulsification, the droplet formation rate is orders of magnitude higher than in spontaneous emulsification methods, and this is reflected in the estimated surface areas and volumes of up-scaled microfluidic devices (see Table 6.2). However, the emulsions prepared by the latter methods are more monodisperse, and that is expected to be an important 'selling point' for microchannel technology for specialty products. Apart from that, the required surface areas for microfluidic devices are rather high compared to membranes and microsieves, and this makes them not immediately suited for large-scale production of low-added value products, given their cost of manufacture.

Most of the microfluidic devices used in literature are made of either silicon or glass. However, some have been produced in polymer via various methods (e.g. Martynova *et al.*, 1997; Roberts *et al.*, 1997; Duffy *et al.*, 1998; Liu *et al.*, 2004; Narasimhan and Papautsky, 2004; Vogelaar *et al.*, 2005), enabling mass production and hence mass parallelization of the devices. It is expected that this will bring down the price of micro devices, which, together with ease of operation and scalability, will be important in the further development of emulsification with micro devices. Please note, that also for these devices interactions with the wall are just as relevant as for silicon based microstructures, and it is expected that also for the polymeric devices surface modification methods need to be developed.

6.5.1. Other products

As mentioned, because of their mildness, microfluidic processes are well suited for products that are sensitive to shear such as e.g. double emulsions (Van der Graaf *et al.*, 2005; a good review is the one by Muschiolik (2007)). The production of capsules produced by layer-by-layer adsorption (Sagis *et al.*, 2008), for which a monodisperse starting emulsion is needed to warrant uniform loading of the capsules would also benefit from the monodisperse emulsions that can be produced by microfluidic technology. Surprisingly, also good results for double emulsions are reported for pre-mix emulsification by Shima *et al.* (2004). Vladisavljevic and coworkers (2006) show that pre-mix emulsification can effectively be used to reduce the size of the droplets while also maintaining high encapsulation efficiency (>90%). For more information on the preparation of multiple emulsions, including membrane emulsification, please see the overview paper by Muschiolik (2007).

The basics presented in this chapter for single emulsions can be applied for various other products, such as polymeric beads (solid and hollow). E.g. premix membrane emulsification has been used for the production of (biodegradable) particles, and in this case, the various time scales that play a role during droplet formation need to be tuned very carefully (Sawalha *et al.*, 2008a). A polymer solution (sometimes the monomer is used which is polymerized later) is mixed with another liquid in which the solvent can dissolve readily, but the polymer cannot. During emulsification, phase separation sets in and the polymer starts to solidify, and this should not take place faster than the actual emulsification process, otherwise the membrane will block. Some recent examples can be found in the work of Zhou and co-workers (2008) who prepared agarose beads of around 5 micron, and Lv and co-workers (2009) who reported chitosan beads which are in the nanometre range.

From solid polymeric beads, it is only a small step to hollow particles, which can be applied as ultrasound contrast agents. In this case, an additional liquid is added to the polymer solution, and this liquid is chosen so that it is not soluble in the continuous phase and remains inside the solidifying polymer droplet. Later this liquid can be removed e.g. by freeze-drying. For the specific case of polylactide, much information was gathered by Sawalha *et al.*, and a scaling relation was reported (2008b), which relates e.g. number of passes, interfacial tension, and applied pressure to the size of the particles. Kooiman and coworkers (2009) demonstrated that it is even possible to load hollow particles with an oil-soluble drug if two extra liquids are added to the polymer solution prior to premix emulsification. One of the liquids can be removed by freeze-drying, and forms the hollow core of the particle that can be activated by ultrasound, while the other liquid contains the drug that initially remains inside the particle, but can be released at increased ultrasound levels at which the bubble bursts. For an overview of the many different products made by membrane/microfluidic emulsification, please consult the review of Vladisavljevic and Williams (2005; Figure 1 in that paper).

6.5.2. Interfacial dynamics

In the various equations presented in this chapter, the interfacial tension is used. The appropriate value during emulsification is not generally known. Either the equilibrium interfacial tension or the interfacial tension of a system without surfactant can be used to determine the range within which the droplet size is expected. However, estimating interfacial tension in these highly dynamic systems is far from trivial. In 1998, the importance of the dynamic interfacial tension was investigated by Schröder and co-workers through the bursting membrane technique, and it could be concluded that for conditions typical for membrane emulsification, a dynamic interfacial tension value is needed. One of the few works we know of that used dynamic interfacial tension values is by Van der Graaf and coworkers (2004), where droplet volume tensiometry was used to extrapolate interfacial tension values to the conditions applied for droplet formation on a microsieve. Steegmans and co-workers (2009c) used a Y-shaped junction and were able to estimate interfacial tensions on time scales that are relevant for any emulsification method. This may well give a refreshing perspective on scaling studies.

In a way, it is not surprising that information on dynamic interfacial tension values is not available, given the lack of methods that allow analysis on the typical time scales relevant to membrane emulsification, which are orders of magnitude shorter than in standard interfacial tension measurement devices.

6.5.3. Computer modelling

Various computational efforts have been made to understand emulsification better. In 2001, Abrahamse and co-workers reported their first results obtained by CFD, and could, for example, show that the interfacial tension and the surface properties, as reflected in the contact angle, are very relevant for the droplet size. Rayner and co-workers (2004) used the Surface Evolver (a package calculating the interfacial shape under static conditions) under conditions for which the force balance does not hold, and they could predict the droplet size with an average error of 8%, which is an improvement compared to using the force balance only. Within the group of Drioli, the force balance models were adjusted to also include amongst others, contact angle effects (De Luca and Drioli, 2006), and further torque balance models have been suggested in recent work of De Luca and co-workers (2008), but unfortunately this did not lead to accurate predictions of the droplet size.

Although the modelling results are once again a step forward in understanding membrane emulsification, the questions regarding the actual mechanism remain (also discussed in the micro technology section). Connected to this, the influence of the (dynamic) interfacial tension on droplet formation is still uncharted territory, and extremely difficult to cover in a computer model.

6.5.4. Design options

Fouling is a serious issue in microfluidic devices: (partial) blockage of the relatively small channels will occur rather easily, and because of this, not all units can be operated simultaneously, or may even start generating droplets of different sizes. On a seemingly lesser scale, wettability of the devices will be influenced by adsorbing species, and especially in small channels that are required to make small droplets, this will have a great effect, and may even cause emulsification to be no longer possible as is also described for membranes (e.g. Schroën *et al.*, 1993). If the to-be-dispersed phase starts wetting the main channel of e.g. a T-junction, the whole structure will be blocked. Therefore, the research dedicated to surface modification is of great importance for further development of microfluidic devices. For some silicon-based materials, appropriate modification methods are available (Arafat *et al.*, 2004 and 2007, or Rosso *et al.*, 2009).

If surface modification does not give the desired results, alternative operation may be needed. E.g. in the group of prof. Mugele (University of Twente, the Netherlands), electro-wetting is used to make droplets rather independent of the flow conditions (e.g. Gu *et al.*, 2008; Mallogi *et al.*, 2008). Also, the glass bead system presented by Van der Zwan and co-workers (2008) is an interesting alternative. It uses pre-mix emulsification by passage through a layer of glass beads instead of a membrane, and can easily be cleaned by re-suspension of the particles, re-assembled, and re-used. Also new (hybrid) developments such as the valve system presented by Abate and co-workers (2009) for flow focusing devices could be interesting for scale-up given the formation frequencies that are in the kilohertz range. These are only a few examples, but it is clear that there are many options waiting to be applied for emulsion production, and for related products.

6.5.5. Concluding remark

Currently, emulsification is still carried out with classic technology for which the relevant design equations are given in the first section of this chapter. Microfluidic technologies are developing fast, but are not in the market yet. Design estimations can be made using Section 6.4 of this chapter. It is expected that especially for low volume/high added-value products microfluidic emulsification will become relevant soon, due to the mildness of the methods, which allows a wider choice of materials, and especially due to the monodispersity of the products.

Acknowledgement

The authors would like to thank prof. Tiny van Boekel and dr. Anita Linnemann for their constructive comments that have helped us make this chapter more complete.

References

Abate, A.R., Romanowsky, M.B., Agresti, J.J. and Weitz, D.A. (2009). Valve-based flow focusing for drop formation. Applied Physics Letters 94: 023503.
Abrahamse, A.J., Van Lierop, R., Van der Sman, R.G.M., Van der Padt, A. and Boom, R.M. (2002) Analysis of droplet formation and interactions during cross-flow membrane emulsification. Journal of Membrane Science 204(1-2): 125-137.

Abrahamse, A.J., Van der Padt, A., Boom, R.M. and De Heij, W.B.C. (2001). Process Fundamentals of Membrane Emulsification: Simulation with CFD. American Institute of Chemical Engineers Journal 47: 1285-1291.

Anna, S.L., Bontoux, N. and Stone, H.A. (2003) Formation of dispersions using "flow focusing" in microchannels. Applied Physics Letters 82(3): 364-366.

Arafat A., Schroën, K., De Smet, L., Sudhölter, E. and Zuilhof, H. (2004) Tailor-made Functionalization of Silicon Nitride Surfaces. Journal of the American Chemical Society 126: 8600-8601.

Arafat, A., Giesbers, M., Rosso, M., Sudhölter, E.J.R., Schroën, C.G.P.H., White, R.G., Yang, L., Linford, M.R. and Zuilhof, H. (2007) Covalent Biofunctionalization of Silicon Nitride Surfaces. Langmuir 23: 6233-6244.

Arbuckle, W.S. (1986) Emulsification. In: Hall, C.W., Farral, A.W. and Rippen, A.L. (ed.), Encyclopaedia of Food Engineering, Avi Pbl. Company Inc., Westport, CT, USA, pp. 286-288.

Aryantia, N., Williams, R.A., Houa, R. and Vladisavljevic G.T. (2006) Performance of Rotating Membrane Emulsification for o/w Production. Desalination 200: 572-574.

Becher, P. (ed.) 1996. Encyclopedia of Emulsion Technology. Vol 1-4. Marcel Dekker, New York, NY, USA.

Behrend, O. and Schubert, H. (2001). Influence of Hydrostatic Pressure and Gas Content on Continuous Ultrasound Emulsification. Ultrasonics Sonochemistry 8: 271-276.

Brennan, J.G. (1986) Emulsification, Mechanical Procedures. In: Hall, C.W., Farral, A.W. and Rippen, A.L. (eds.), Encyclopaedia of Food Engineering. Avi Pbl. Company Inc., Westport, CT, USA, pp. 288-291.

Charcosset, C. (2009). Preparation of Emulsions and Particles by Membrane Emulsification for the Food Processing Industry. Journal of Food Engineering 92: 241-249.

Charcosset, C., Limayem, I. and Fessi, H. (2004) The Membrane Emulsification Process – a Review. Journal of Chemical Technology and Biotechnology 79: 209-218.

De Luca, G., Di Maio, F.P., Di Renzo A. and Drioli, E. (2008) Droplet Detachment in Cross-flow Membrane Emulsification: Comparison among Torque- and Force-based Models. Chemical Engineering and Processing 47: 1150-1158.

De Luca, G. and Drioli, E. (2006) Force Balance Conditions for Droplet Formation in Cross-flow Membrane Emulsifications. Journal of Colloid and Interface Science 294: 436-448.

Duffy, D.C., McDonald, J.C., Schueller, O.J.A. and Whitesides, G.M. (1998) Rapid Prototyping of Microfluidic Systems in Poly(dimethylsiloxane). Analytical Chemistry 70: 4974-4984.

Eisner, V. (2007) Emulsion Processing with a Rotating Membrane (ROME). Dissertation number 17153, ETH Zürich, Switzerland.

Gijsbertsen-Abrahamse, A.J., Van der Padt, A. and Boom, R.M. (2004) Status of Cross-flow Membrane Emulsification and Outlook for Industrial Application. Journal of Membrane Science 230: 149-159.

Gijsbertsen-Abrahamse, A.J. 2003. Membrane emulsification: process principles. PhD thesis, Wageningen University, Wageningen, the Netherlands.

Gu, H., Malloggi, F., Vanapalli S.A. and Mugele, F. (2008) Electrowetting-enhanced microfluidic device for drop generation. Applied Physics Letters 93: 183507.

Guzey, D. and McClements D.J. (2006) Formation, stability and properties of multilayer emulsions for application in the food industry. Advances in Colloid and Interface Science 128-130: 227-248.

Hiemenz, P.C. (1986) Principles of colloid and surface chemistry. Marcel Dekker, Inc., New York, NY, USA.

Joscelyne, S.M. and Trägårdh, G. (2000) Membrane Emulsification – a Literature Review. Journal of Membrane Science 169: 107-117.

Karbstein, H. and Schubert, H. (1995). Developments in the Continuous Mechanical Production of Oil-in-water Macro-emulsions. Chemical Engineering and Processing 34: 205-211.

Katoh, R., Asano, Y. Furuya, A. Sotoyama, K. and Tomita, M. (1996) Preparation of Food Emulsions using a Membrane Emulsification System. Journal of Membrane Science 113: 131-135.

Kawakatsu, T., Kikuchi Y. and Nakajima M. (1997) Regular-sized Cell Creation in Microchannel Emulsification by Visual Microprocessing Method. Journal of the American Oil Chemists' Society 74: 317-321.

Kikuchi, Y., Sato, K., Ohki, H. and Kaneko, T. (1992) Optically Accessible Microchannels Formed in a Single-Crystal Silicon Substrate for Studies of Blood Rheology. Microvascular Research 4: 226-240.

Kobayashi, I., Lou, X. F., Mukataka, S. and Nakajima M. (2005a) Preparation of Monodisperse Water-in-oil-in-water Emulsions using Microfluidization and Straight-through Microchannel Emulsification. Journal of the American Oil Chemists Society 82: 65-71.

Kobayashi, I., Mukataka, S. and Nakajima, M. (2004) CFD Simulation and Analysis of Emulsion Droplet Formation from Straight-through Microchannels. Langmuir 20: 9868-9877.

Kobayashi, I., Mukataka, S. and Nakajima, M. (2005b) Novel Asymmetric Through-hole Array Microfabricated on a Silicon Plate for Formulating Monodisperse Emulsions. Langmuir 21: 7629-7632.

Kobayashi, I., Nakajima, M., Chun, K., Kikuchi, Y. and Fujita, H. (2002) Silicon Array of Elongated Through-holes for Monodisperse Emulsion Droplets. American Institute of Chemical Engineers Journal 48: 1639-1644.

Kobayashi, I., Uemura, K. and Nakajima, M. (2006) CFD Study of the Effect of a Fluid Flow in a Channel on Generation of Oil in-water Emulsion Droplets in Straight-through Microchannel Emulsification. Journal of Chemical Engineering of Japan 39: 855-863.

Kooiman, K., Böhmer, M.R., Emmer, M., Vos, H.J., Chlon, C., Shi, W.T., Hall, C.S., De Winter, S.H.P.M., Schroën, K., Versluis, M., De Jong, N. and Van Wamel, A. (2009) Oil-filled Polymer Microcapsules for Ultrasound-mediated Delivery of Lipophilic Drugs. Journal of Controlled Release 133: 109-118.

Krog, N.J., Riisom, T.H. and Larson, K. (1985) Applications in food industry. In: Becher, P. (ed.) Encyclopedia of Emulsion Technology. Vol 2. Applications. Marcel Dekker, New York, NY, USA, pp: 58-127.

Lambrich, U. and Schubert, H. (2005). Emulsification using Microporous Systems. Journal of Membrane Science 257: 76-84.

Link, D.R., Anna, S.L., Weitz, D.A. and Stone, H.A. (2004) Geometrically mediated break-up of drops in microfluidic devices. Physical review letters, 92(5): 054503/1-054503/4.

Liu, H., Nakajima, M. and Kimura, T. (2004) Production of Monodispersed Water-in-Oil Emulsions Using Polymer Microchannels. Journal of the American Oil Chemists' Society 81: 705-711.

Lucassen-Reynders, E.H. (1996) Dynamic interfacial properties in emulsification. In: Becher, P. (ed.) Encyclopedia of Emulsion Technology. Vol 4. Marcel Dekker, New York, NY, USA, pp. 63-90.

Lv, P-P., W. Wei, F.-L. Gong, Y.-L. Zhang, H-Y. Zhao, J-D. Lei, Wang, L.-Y. and Guang-Hui, M.A. (2009) Preparation of Uniformly Sized Chitosan Nanospheres by a Premix Membrane Emulsification Technique. Industrial & Engineering Chemistry Research 48: 8819-8828.

Lyklema, J. (1991) Fundamentals of Interface and Colloid Science. Academic Press Limited, London, UK.

Malloggi, F., Gu, H., Banpurkar, A.G., Vanapalli, S.A. and Mugele, F. (2008) Electrowetting - A versatile tool for controlling microdrop generation The European Physical Journal E 26: 91-96.

Martynova, L., Locascio, L.E., Gaitan, M., Kramer, G.W, Christensen, R.G. and MacCrehan, W.A. (1997) Fabrication of Plastic Microfluid Channels by Imprinting Methods. Analytical Chemistry 69: 4783-4789.

Muschiolik, G. (2007) Multiple Emulsions for Food Use. Current Opinion in Colloid & Interface Science 12: 213-220.

Nakashima, T. and Shimizu, M. (1986) Porous Glass from Calcium Alumino Boro-silicate Glass. Ceramics Japan 21: 408.

Nakashima, T., Shimizu M. and Kukizaki, M. (1991) Membrane Emulsification by Microporous Glass. Key Engineering Materials 61-62: 513.

Narasimhan, J. and Papautsky, I. (2004) Polymer Embossing Tools for Rapid Prototyping of Plastic Microfluidic Devices. Journal of Micromechanics and Microengineering 14: 96-103.

Nazir, A., Schroën, K. and Boom, R. (2010). Pre-mix emulsification: A review. Journal of Membrane Science 362, 1-2: 1-11.

Nisisako, T. and Torii, T. (2008) Microfluidic Large-scale Integration on a Chip for Mass Production of Monodisperse Droplets and Particles. Lab On A Chip 8: 287-293.

Peng, S.J. and Williams, R.A. (1998) Controlled Production of Emulsions using a Crossflow Membrane. Part I: Droplet Formation from a Single Pore. Trans IChemE, 76: 894-901.

Rayner, M. and Trägårdh, G. (2002) Membrane Emulsification Modelling: How can We get from Characterisation to Design? De salination 145: 165-172.

Rayner, M., Trägårdh, G., Trägårdh, C. and Dejmek, P. (2004) Using the Surface Evolver to Model Droplet Formation Processes in Membrane Emulsification. Journal of Colloid and Interface Science 279: 175-185.

Roberts, M.A., Rossier, J.S, Bercier, P. and Girault, H. (1997) UV Laser Machined Polymer Substrates for the Development of Microdiagnostic Systems. Analytical Chemistry 69: 2035-2042.

Rosso, M., Giesbers, M., Arafat, A., Schroën, K. and Zuilhof, H. 2009. Covalently attached Organic Monolayers on SiC and SixN4 Surfaces: Formation using UV Light at Room Temperature. Langmuir 25: 2172-2180.

Sagis, L.M.C., De Ruiter, R., Rossier, J., De Ruiter, M.J., Schroën, K., Van Aelst, A.C., Kieft, H., Boom, R.M. and Van der Linden, E. (2008) Polymer Microcapsules with a Fiber-Reinforced Nanocomposite Shell. Langmuir 24: 1608-1612.

Sawalha, H.I.M. (2009) Polylactide Microcapsules and Films: Preparation and Properties. PhD thesis, Wageningen University, Wageningen, the Netherlands.

Sawalha, H., Fan, Y., Schroën K. and Boom, R. (2008b). Preparation of Hollow Polylactide Microcapsules through Premix Membrane Emulsfication - Effects of Nonsolvent Properties. Journal of membrane science 325: 665-671.

Sawalha, H., Purwanti, N., Rinzema, A., Schroën, K. and Boom, R. (2008a). Polylactide Microspheres prepared by Premix Membrane Emulsification - Effects of Solvent Removal Rate. Journal of membrane science 310: 484-493.

Schadler, V. and Windhab, E.J. (2006) Continuous Membrane Emulsfication by using a Membrane System with Controlled Pore Distance. Desalination 189: 130-135.

Schröder, V., Behrend, O. and Schubert, H. (1998) Effect of dynamic interfacial tension on the emulsification process using microporous, ceramic membranes. Journal of Colloid Interface Science 202: 334-340.

Schröder, V. and Schubert H. (1999) Production of Emulsions using Microporous, Ceramic Membranes. Colloids and Surfaces A: Physicochemical and Engineering Aspects 152: 103-109.

Schroën, C.G.P.H., Wijers, M.C., Cohen-Stuart, M.A., Van der Padt, A. and Van 't Riet, K. (1993) Membrane Modification to avoid Wettability Changes due to Protein Adsorption in an Emulsion/membrane Bioreactor. Journal of Membrane Science 80: 265-274.

Schubert, H. and Armbruster, H. (1992) Principles of formation and stability of emulsions. International Journal of Chemical Engineering 32: 14.

Shima, M., Kobayashi, Y., Fujii, T., Tanaka, M., Kimura, Y., Adachi, S. and Matsuno, R. (2004) Preparation of fine W/O/W Emulsion through Membrane Filtration of Coarse W/O/W Emulsion and Disappearance of the Inclusion of outer Phase Solution. Food Hydrocolloids 18: 61-70.

Smulders, P.E.A. (2000) Formation and stability of emulsions made with proteins and peptides. PhD thesis, Wageningen University, Wageningen, the Netherlands.

Steegmans, M.L.J. (2009) Emulsification in microfluidic Y- and T-junctions. PhD thesis, Wageningen University. ISBN: 978-90-8585-457-9.

Steegmans, M.L.J., Schroën C.G.P.H. and Boom, R.M. (2009a) Characterization of Emulsification at Flat Microchannel Y Junctions. Langmuir 25: 3396-3401.

Steegmans, M.L.J., Schroën, C.G.P.H. and Boom, R.M. (2009b) Generalised insights in droplet formation at T-junctions through statistical analysis. Chemical Engineering Science 64 (13): 3042-3050.

Steegmans, M.L.J., Schroën, C.G.P.H. and Boom, R.M. (2009c) Dynamic Interfacial Tension Measurements with Microfluidic Y-Junctions. Langmuir 25 (17): 9751-9758.

Steegmans, M.L.J., Schroën, C.G.P.H. and Boom, R.M. (2010) Microfluidic Y-Junctions: A Robust Emulsification System with Regard to Junction Design. AIChE Journal 56(7): 1946-1949.

Stillwell, M.T., Holdich, R.G., Kosvintsev, S.R., Gasparini, G. and Cumming, I.W. (2007) Stirred Cell Membrane Emulsification and Factors Influencing Dispersion Drop Size and Uniformity. Industrial & Engineering Chemistry Research 46: 965-972.

Sugiura, S., Nakajima, M. Iwamoto, S. and Seki, M. (2001) Interfacial Tension driven Monodispersed Droplet Formation from Microfabricated Channel Array. Langmuir 17: 5562-5566.

Sugiura, S., Nakajima, M., Kumazawa, N., Iwamoto, S. and Seki, M. (2002). Characterization of spontaneous Transformation-based Droplet Formation during Microchannel Emulsification. Journal of Physical Chemistry B 106: 9405-9409.

Sugiura, S., Nakajima, M. and Seki, M. (2002) Preparation of Monodispersed Emulsion with Large Droplets using Microchannel Emulsification. Journal of the American Oil Chemists Society 79: 515-519.

Sugiura, S., Nakajima, M., Ushijima, H., Yamamoto, K. and M. Seki. 2001. Preparation Characteristics of Monodispersed Water-in-oil Emulsions using Microchannel Emulsification. Journal of Chemical Engineering Japan 34: 757-765.

Sugiura, S., Nakajima, M., Yamamoto, K., Iwamoto, S., Oda, T., Satake, M. and Seki, M. (2004) Preparation Characteristics of Water-in-oil-in-water Multiple Emulsions using Microchannel Emulsification. Journal of Colloid and Interface Science 270: 221-228.

Suzuki, K., Hayakawa, K. and Hagura, Y. (1999) Preparation of High Concentration O/W and W/O Emulsions by the Membrane Phase Inversion Emulsification Using PTFE Membranes. Food Science and Technology Research 5: 234-238.

Van der Graaf, S., Nisisako, T., Schroën, C.G.P.H., Van der Sman R.G.M. and Boom R.M. (2006) Lattice Boltzmann Simulations of Droplet Formation in a T-shaped Microchannel. Langmuir 22: 4144-4152.

Van der Graaf, S., Schroën, C.G.P.H., Van der Sman, R.G.M. and Boom R.M. (2004). Influence of Dynamic Interfacial Tension on Droplet Formation during Membrane Emulsification. Journal of Colloid and Interface Science 277: 456-463.

Van der Graaf, S., Schroën, C.G.P.H. and Boom, R.M. (2005a) Preparation of Double Emulsions by Membrane Emulsification—a Review. Journal of Membrane Science 251: 7-15.

Van der Graaf, S., Steegmans, M.L.J., Van der Sman, R.G.M., Schroën, C.G.P.H. and Boom, R.M. (2005b) Droplet Formation in a T-shaped Microchannel Junction: A Model System for Membrane Emulsification. Colloids and Surfaces A: Physicochemical Engineering Aspects 266: 106-116.

Van der Zwan, E. (2008) Emulsification with microstructured systems. Process principles. PhD thesis Wageningen University, Wageningen, the Netherlands.

Van der Zwan, E., Schroën, C.G.P.H., Van Dijke, K. and Boom, R.M. (2006) Visualization of Droplet Break-up in Pre-mix Membrane Emulsification using Microfluidic Devices. Colloids and Surfaces A-Physicochemical and Engineering Aspects 277: 223-229.

Van der Zwan, E., Schroën, C.G.P.H. and Boom, R.M. (2008) Pre-mix Membrane Emulsification by Using a Packed Layer of Glass Beads. American Institute of Chemical Engineers Journal 54: 2190-2197.

Van der Zwan, E., Schroën, C.G.P.H. and Boom, R.M. (2009a) A Geometric Model for the Dynamics of Microchannel Emulsification. Langmuir 25: 7320-7327.

Van der Zwan, E., Van der Sman, R. Schroën, K. and Boom, R.M. (2009b) Lattice Boltzmann Simulations of Droplet Formation during Microchannel Emulsification. Journal of Colloid and Interface Science 335: 112-122.

Van Dijke, K.C. 2009. Emulsification with microstructures. PhD thesis, Wageningen University, ISBN: 978-90-8585-495-1.

Van Dijke K.C., K. Schroën and R.M. Boom. 2008. Microchannel Emulsification: From Computational Fluid Dynamics to Predictive Analytical Model. Langmuir. 24: 10107-10115.

Van Dijke, K., Kobayashi, I., Schroën, K., Uemura, K., Nakajima, M. and Boom, R. (2010a) Effect of viscosities of dispersed and continuous phases in microchannel oil-in-water emulsification. Microfluidics Nanofluidics 9: 77-85.

Van Dijke, K, De Ruiter, R., Schroën, K. and Boom, R. (2010 b) The mechanism of droplet formation in microfluidic EDGE systems. Soft Matter 6: 321-330.

Van Dijke, K.C., Schroën, C.G.P.H., Van der Padt, A. and Boom, R.M. (2010c) EDGE emulsification for food-grade dispersions. Journal of Food Engineering 97(3): 348-354.

Van Dijke, K.C., Veldhuis, G., Schroën, C.G.P.H. and Boom, R.M. (2009) Parallelized edge-based droplet generation (EDGE) devices. Lab On a Chip 9: 2824-2830.

Van Rijn, C.J.M. (2004) Nano and micro engineered membrane technology. Membrane Science and Technology Series 10. Elsevier, 384 p.

Van Rijn, C.J.M. and Elwenspoek, M.C. (1995) Micro Filtration Membrane Sieve with Silicon Micro Machining for Industrial and Biomedical Applications. Proceedings Institute of Electrical and Electronics Engineers 29: 83-87.

Van Steijn, V., Kleijn C.R. and Kreutzer M.T. (2010) Predictive model for the size of bubbles and droplets created in microfluidic T-junctions. Lab Chip 10: 2513-2518.

Vladisavljevic, G.T. and Schubert, H. (2003) Preparation of Emulsions with a Narrow Particle Size Distribution Using Microporous -Alumina Membranes. Journal of Dispersion Science and Technology 24: 811-819.

Vladisavljevic, G.T., Shimizu, M. and Nakashima, T. (2006) Production of Multiple Emulsions for Drug Delivery Systems by repeated SPG Membrane Homogenization: Influence of Mean Pore Size, Interfacial Tension and Continuous Phase Viscosity. Journal of Membrane Science 284: 373-383.

Vladisavljevic, G.T., Tesch S. and Schubert, H. (2002) Preparation of Water-in-oil Emulsions using Microporous Polypropylene Hollow Fibers: Influence of some Operating Parameters on Droplet Size Distribution. Chemical Engineering and Processing 41: 231-238.

Vladisavljevic, G.T. and Williams, R.A. (2005) Recent Developments in Manufacturing Emulsions and Particulate Products using Membranes. Advances in Colloid and Interface Science 113: 1-20.

Vladisavljevic, G.T., Kobayashi, I. and Nakajima, M. (2010) Effect of dispersed phase viscosity on maximum droplet generation frequency in microchannel emulsification using asymmetric straight-through channels. Microfluidics, Nanofluidics, DOI: 10.1007/s10404-010-0750-9.

Vogelaar, L., Lammertink, R.G.H., Barsema, J.N., Nijdam, W., Bolhuis-Versteeg, L.A.M., Van Rijn, C.J.M. and Wesseling, M. (2005) Phase Separation Micromolding: a new Generic Approach for Microstructuring various Materials. Small 1: 645-655.

Walstra, P. (1983) Formation of Emulsions. In: Becher P. (ed.) Encyclopedia of Emulsion Technology. Vol 1. Basic aspects. Marcel Dekker, New York, NY, USA, pp. 58-127.

Walstra, P. (1996) Emulsion stability. In: Becher P. (ed.) Encyclopedia of Emulsion Technology. Vol 4. Marcel Dekker, New York, NY, USA, pp. 58-127.

Walstra, P. (2003) Physical chemistry of Foods. Marcel Dekker inc., New York, NY, USA.

Walstra, P., Wouters, J.T.M. and Geurts, T.J. (2006). Dairy Science and Technology. Taylor & Francis, Boca Raton, FL, USA.

Walstra, P. and Smulders, P.E.A. (1998) Emulsion formation. In: Binks B.P. (ed.) Modern Aspects of Emulsion Science. Royal Society of Chemistry Cambridge: 56-99.

Yuan, Q., Houa, R., Aryantia, N., Williams, R.A., Biggs, S., Lawson, S., Silgram, H., Sarkar M. and Birch, R. (2008). Manufacture of Controlled Emulsions and Particulates using Membrane Emulsification. Desalination 224: 215-220.

Zhou, Q-Z., Wang, L-Y., Ma, G-H. and Su, Z-G. (2008) Multi-stage Premix Membrane Emulsification for Preparation of Agarose Microbeads with Uniform Size. Journal of Membrane Science 322: 98-104.

7. Food packaging design[1]

Matthijs Dekker
Product Design and Quality Management, Wageningen University

7.1. Introduction

Packaging design is an essential part of product design. Packaging plays an important role in the quality perception of food products. A suitable design of packaging can maintain a good internal quality of the food product and it can also add certain additional quality aspects to food products. The functions of food packaging can be divided into three categories:

Product
* Protection
* Stabilisation

Logistics
* Distribution

Consumer
* Image
* Information
* Portioning
* Integrity
* Convenience

Historically the main emphasis has been on the stabilisation and protection of foods. Large-scale packaging was initiated by Nicolas Appert with the 'invention' of sterilising food products that were hermetically sealed from the environment in glass jars and later in metal cans.

The stabilisation and protection of our foods are still the main functions of packaging nowadays, but the logistic and consumer-oriented functions have become increasingly important. The marketing aspects of packaging play a particularly dominant role in the design of new packaging. On average some 10% of the cost of food products is spent on packaging. For new special designs this percentage can be much higher, indicating that the package should really add significant benefits for the potential buyer. Increasing shelf life, product safety, image and convenience are important features of food packaging.

---

[1] This chapter is an extended and updated version of the chapter 'Developments in food packaging' by M. Dekker in 'Innovation in agri-food systems; Product quality and consumer acceptance', edited by W.M.F. Jongen and M.T.G. Meulenberg, Wageningen Academic Publishers, 2005.

Consumer trends, such as the demand for less intensely processed foods containing less or no preservatives, make the stabilising and protective function of packaging even more crucial. Other trends are the demand for more convenience, another aspect in which the packaging can play an important role. The rest of this chapter will highlight the basic aspects of dealing with the barrier properties of packaging materials for food, and some novel technological developments in relation to dealing with recent consumer trends. An extensive description of the principles of food packaging, material properties and packaging applications to a wide variety of food products has been described by Robertson (2005).

7.2. Packaging materials as a barrier

One of the basic functions of food packaging materials is to act as a barrier for gases, water and aromas. The kinetics of food quality decay and food safety is highly dependent on the composition of the atmosphere inside a package. Therefore, the shelf life of food products is strongly dependent on the barrier properties of the packaging materials. The main compounds of interest in this respect are: oxygen, carbon dioxide, water, nitrogen and product-specific volatiles. Transfer of these compounds through packaging materials can be described by a combination of Fick's law for diffusion:

$$\frac{\varphi}{A} = D \cdot \frac{dC}{dx} \tag{7.1}$$

in which φ is the flux through the packaging material in mol/s, A is the surface area in m^2, D is the diffusion coefficient of the compound in m^2/s, C is the compound concentration in mol/m^3 and x is the distance in m.

Since the limiting diffusion of the compounds is generally determined by the barrier properties of the packaging material the values of A, D, C and x refer to the packaging material. The steady state diffusion through the material can then be described by:

$$\frac{\varphi}{A} = D \cdot \frac{(C_1 - C_2)}{x} \tag{7.2}$$

in which the subscripts 1 and 2 refer to both sides of the packaging material.

In order to estimate the concentrations of the compounds at the sides of the material one can use Henry's law for the partitioning of compounds between gas and the packaging material:

$$C_i = S \cdot p_i \tag{7.3}$$

in which S is the solubility coefficient in $(mol \cdot s^2)/(m^3 \cdot kg)$ and p_i the partial pressure of the compound in the surrounding gas (internal or external atmosphere) in Pa.

Combining Equation (7.1) and (7.2) yields:

$$\frac{\varphi}{A} = D \cdot S \cdot \frac{(p_1 - p_2)}{x}$$ (7.4)

In this equation the term $D{\cdot}S$ is often replaced by the permeability coefficient P of the packaging material. This coefficient can be obtained from experimentally-determined fluxes at a certain difference in partial pressures through a specified surface area at known thickness. The value for P is dependent on external conditions like temperature and humidity.

In practical applications of packaging materials, instead of specifying the permeability of a material for a certain compound, on many occasions the gas transmission rate (GTR) and the water vapour transmission rate (WVTR) are given for a certain film thickness (usually 25 µm), temperature and humidity. Units for GTR's are generally: ml/(day·m²·bar), and for WVTR's generally g/(day·m²). Since the values of GTR and WVTR are given at a reference thickness they have to be converted to their value at the actual thickness of the material under consideration. This can be done by using Equation (7.5):

$$GTR_a = GTR_{ref} \cdot \frac{l_{ref}}{l_a}$$ (7.5)

in which l is the thickness of the material and the subscript a refers to 'actual' and ref to 'reference'. For WVTR values the GTR value in Equation (7.5) can simply be replaced by the WVTR values.

Calculating fluxes of compounds from these converted values of GTR and WVTR can easily be done with Equation (7.6) and (7.7):

$$\varphi_j = GTR_j \cdot A \cdot (p_{j,1} - p_{j,2})$$ (7.6)

$$\varphi_w = WVTR \cdot A \cdot (RH_1 - RH_2)$$ (7.7)

in which the subscript j refers to the type of compound, subscript w refers to water and RH to relative humidity (expressed as a fraction of 100%). The units of φ, A and p depend on the units in which the GTR and WVTR are given.

In many applications not a single polymeric material is used, but rather laminates of different materials. To estimate the GTR or WVTR of these laminated or coated materials one can use Equation (7.8), which gives the total resistance (expressed as 1/GTR) of the laminate as a function of the individual resistances of all layers:

$$\frac{1}{GTR_{laminate}} = \sum_{i=1}^{n} \frac{1}{GTR_i}$$ (7.8)

in this equation i refers to the number of the layer in a total of n layers, the values for GTR of the layers have to be converted to the actual thickness by using Equation 7.5 first.

For materials such as glass and metal packaging transfer of these compounds can usually be assumed to be zero due to the extremely low diffusion rates in these materials. For polymeric materials, however, diffusion should be taken into account.

7.3. Modified atmosphere packaging

An important technique for reducing quality change and enhancing the shelf life of food products is to replace the air around the product with another gas mixture. In this way undesired reactions like microbial growth, oxidation, physiological ageing and ripening can be slowed down. Two basic types of this Modified Atmosphere Packaging (MAP) technology can be distinguished. The first type, also called 'gas packaging' is meant for products that are not physiologically active, like ready meals, nuts, meat, fish, etc. It consists of the removal of air during the packaging process by either vacuum formation followed by gas addition or by replacing air with a constant flow of gas (flow packing). To maintain the gas atmosphere the resistance towards gas permeation of the package should be high. Gases used are mainly CO_2 and N_2 combinations; O_2 is usually not desired for the product (Devlieghere et al., 2003). Meat (red) is an exception; to maintain the appealing red colour oxygen is required within the headspace of these products. Atmospheres containing low levels of CO have also been suggested for meat (Gill, 2003). Shelf life of products that are gas packaged is enhanced mainly by a reduction of microbial growth and oxidation reactions. The growth of pathogens requires special attention under these conditions (Devlieghere *et al.*, 2003; Gill, 2003). Developments of high O_2 MAP (70-100% O_2) look promising for certain fresh products (Day, 2003). One example is the packaging of cod for which an optimal atmosphere was determined to be 63% oxygen and 37% carbon dioxide (Sivertsvik, 2007).

The second type, also called equilibrium MAP, is used for products, like fresh fruits and vegetables, which are physiologically active. In this case the change in the headspace atmosphere is achieved by the respiration rate of the product itself, in combination with a well-matched permeability property of the packaging material. To maintain an optimal gas atmosphere in this case, a low resistance towards gas permeation of the package is usually required. Depending on the type of fruit and vegetables atmospheres containing typically in the range of 3-10% CO_2 and 3-10% O_2 are optimal for enhancing the shelf life of these products.

The design of equilibrium MAP has to consider both the respiration rate of the product and the transmission rates of the packaging material for the relevant gases. A mathematical approach for this and its application for minimally-processed fruits are given by Del Nobile *et al.* (2007). For highly respiring products micro-perforated packaging films are often used in order to reach the desired fluxes for oxygen and carbon dioxide and to prevent the product from becoming anaerobic.

Both types of MAP discussed are frequently used in packaging designs for new products. Enhanced product stability offers benefits for the consumers, but also greater flexibility and efficiency within the distribution chain.

7.4. Active packaging

Active Packaging is another concept aiming to maintain product quality for longer periods of time. The European Project 'Actipak' (FAIR-Project PL 98-4170) defined the concept as:
'Packaging that changes the condition of the packaged food to extend shelf life or to improve safety or sensory properties, while maintaining the quality of the food.'

Different concepts of active packaging have been developed. Important active concepts are:
• oxygen scavengers;
• ethylene scavengers;
• moisture regulators;
• antimicrobial emitters;
• antioxidant emitters;
• flavour scavengers/emitters.

The active ingredient can be incorporated into the package in the form of a sachet or can be integrated into the film or wall of the package (Ozdemir and Floros, 2004).

In Figure 7.1 an example is given of a sachet-type active package with an oxygen scavenger.

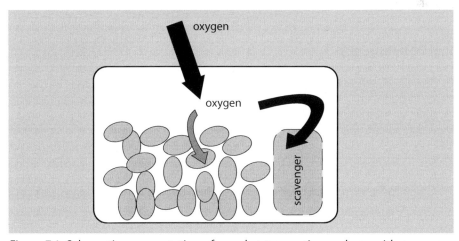

Figure 7.1. Schematic representation of a sachet-type active package with an oxygen scavenger for the protection of an oxidation-sensitive product.

Many interesting results with considerable shelf-life enhancements have been published. Commercialisation has mainly been in the USA and Japan. In Europe many applications have been restricted in the past by the European migration regulations (De Kruijf *et al.*, 2002). European regulation has recently changed to allow for many applications. Some of the possible applications, however, require special consideration with respect to food safety. For instance, the removal of off-flavour from the package of fresh fish may mask spoilage of the product for the consumer. A review of recent developments of active packaging and their application for meat products has been described by Kerry *et al.* (2006).

7.5. High pressure processing

High Pressure Processing (HPP) is a non-thermal food preservation method. The method consists of applying pressures of 300-800 MPa for a limited amount of time (usually several minutes) to packaged or unpackaged food (Toepfl *et al.*, 2006) When packaged foods are treated with this technique the integrity of the packaging material after the treatment is important for the safety and quality of the product. Caner *et al.* (2004) reviewed published research on these aspects and concluded that in general flexible structures respond well to HPP, especially when they are only based on plastics. Only reversible changes occurred during the treatment. Metallised films were greatly affected in their barrier properties due to rupturing because of uneven compressibility of the different layers. Inorganic coating of Al_2O_3 and SiO_2 showed good behaviour.

The relative high price of HPP due to high investment costs has limited its application to some high-value products for which the quality improvements of this technique over conventional thermal treatments are high. Examples are high-quality fruit juices and guacamole.

7.6. Convenience

In packaging design adding convenience benefits for the consumer is always an important aspect. Easy opening, re-closure and dosing without spilling seem to be essential characteristics of packages, yet this can still lead to major frustrations, especially for certain consumer groups like the elderly and children. Technological developments in this area are focussing on adding convenience through creative designs, developing new materials like easy-peal films, but also in the form of adding extra functions to the packaging. One important example of this is the incorporation of susceptor materials into microwavable packages. These metal alloys convert microwave energy into infrared radiation, thereby enabling surface temperatures on the product well over 100 °C, which is essential for producing products with a crispy outer coating. The use of intelligent packaging concepts (see below) can also greatly enhance the convenience factor of foods.

7.7. Intelligent packaging

Intelligent Packaging can be defined as: 'Packaging that monitors or provides information on product, product quality, security or whereabouts during transport, storage, retail and/or use' (www.piranet.com). Although conventional bar-codes could be part of this definition as they provide information on the product identity, more high-tech concepts have been developed in this area. The development of RFID (radio frequency identification) tags for packages in particular provides options for sophisticated applications. Product recognition without the need for scanning in shops and the prevention of product theft are obvious applications of this concept. More creative applications of product recognition can also be envisaged however: ready meals that instruct an oven or microwave on how they are to be prepared, and refrigerators that manage stock control for the consumer are just some examples.

Another important development is linking sensor technology to RFID tags or other devices allowing for quality information to be transmitted to the environment. Time-temperature indicators (TTI's) are an example of this, but sensors for measuring specific compounds that are an indicator for product quality are also being developed. Commercialisation of these concepts has already occurred with temperature dependent labels that give a visual indication to the consumer of possible temperature abuse of the product (See example in Figure 7.2).

A review of recent developments in intelligent packaging and their application for meat products has been described by Kerry *et al.* (2006).

7.8. Modelling in package design

Mathematical modelling for packaging design has been applied in several fields. For food safety an important field is the migration of components from the packaging materials into food (Arvanitoyannis and Bosnea, 2004). For regulatory purposes experimental designs with food simulants are allowed. An emerging field is the application of predictive models for measuring the extent of migration for

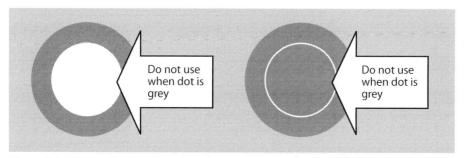

Figure 7.2. Schematic representation of a temperature-sensitive label.

regulatory purposes (Helmroth *et al.*, 2002, 2005 and Stoffers *et al.*, 2005). Two different modelling approaches have been suggested. Stochastic modelling provides insight into the probability of having migration above a certain safety limit set by the regulations (Helmroth *et al.*, 2002; 2005). This information can be used in a risk-assessment calculation. Another approach adopted by European regulation is an empirical approach based on worst-case scenarios (Baner *et al.*, 1996).

With respect to food quality modelling much research has been directed towards the design of equilibrium MAP for respiring products (Hertog *et al.*, 2003). Dekker *et al.* (2002) describe a modelling approach for oxidation sensitive products in different packaging concepts. With respect to the developments in intelligent packaging concepts, such quality change models for food products are needed to be able to translate the information from time-temperature-indicators (TTI's) into information on food quality. Critical reactions within the food have to be identified and characterised with respect to kinetic parameters to be able to predict the quality change as a function of time and temperature.

7.9. Conclusions

For packaging design many functions of the package have to be taken into account simultaneously. Marketing and technological functions have to be integrated into one design. Concepts for shelf-life enhancement of food products can, for many products, be based on modified atmosphere or active packaging concepts. Packages with added functionality will allow for communication of the package with the environment. This can revolutionise the convenience functions built into packages and provide active and online control of the actual food quality and safety. Mathematical modelling of food quality in relation to the package design can help in efficiently designing new packages and packaging materials and is essential for the further development and application of intelligent packaging systems.

References

Arvanitoyannis, I.S. and Bosnea, L. (2004). Migration of Substances from Food Packaging Materials to Foods, Critical Reviews in Food Science and Nutrition, 44:63–76.

Baner, L., Brandsch, J., Franz, R. and Piringer, O.G. (1996). The application of a predictive migration model for evaluation the compliance of plastic materials with European food regulations, Food Additives and Contaminants 13: 587-601.

Caner C., Hernandez, R.J and, Harte B.R. (2004). High-pressure processing effects on the mechanical, barrier and mass transfer properties of food packaging flexible structures: a critical review , Packaging Technology and Science 17(1): 23-29.

Day, B. (2003). Novel MAP Applications for Fresh-Prepared Produce. In: Novel Food Packaging Techniques, Ed. Raija Ahvenainen, Woodhead Publishing Ltd, Cambridge UK.

De Kruijf N.N., van Beest M., Rijk R., Sipilainen-Malm T., Paseiro L.P. and De Meulenaer B. (2002). Active and intelligent packaging: applications and regulatory aspects, Food Addit Contam 19:144-162.

Del Nobile, M.A., Licciardello, F. Scrocco, C., Muratore, G. and Zappa, M. (2007). Design of plastic packages for minimally processed fruits, Journal of Food Engineering 79(1):217-224.

Dekker, M., Kramer, M., van Beest M. and Luning P. (2002). Modeling Oxidative Quality Changes in Several Packaging Concepts, In proceedings Worldpak 2002, Michigan, USA, CRC Press, Boca Raton, Volume 1, pp. 297-303.

Devlieghere, F. , Debevere, J. and Gil, M. (2003). MAP; Product Safety and Nutritional Quality. In: Novel Food Packaging Techniques, R. Ahvenainen (ed.), Woodhead Publishing Ltd, Cambridge UK.

Gill, C. (2003). Active Packaging in Practice: Meat. In: Novel Food Packaging Techniques, R. Ahvenainen (ed.), Woodhead Publishing Ltd, Cambridge UK.

Helmroth, E., Rijk, R., Dekker, M. and Jongen, W.M.F. (2002). Predictive modelling of migration from packaging materials into food products for regulatory purposes, Trends in Food Science and Technology 13:102-109.

Helmroth, E., Varekamp, C. and Dekker, M. (2005.) Stochastic modelling of migration from polyolefins, J. Sci. Fd. Agr. 85: 909-916.

Hertog, M. and Banks, N. (2003). Improving MAP through Modelling, in: Novel Food Packaging Techniques, Ed. Raija Ahvenainen, Woodhead Publishing Ltd, Cambridge UK.

Kerry, J.P., O'Grady, M.N., and Hogan, S.A. (2006). Past, current and potential utilisation of active and intelligent packaging systems for meat and muscle-based products: A review. Meat Science 74:113-130.

Ozdemir, M. and Floros, J.D. (2004). Active food packaging technologies, Critical Reviews in Food Science and Nutrition 44:185-193.

Robertson, G.L. (2005). Food Packaging: Principles and Practice, Second Edition, Taylor & Francis/CRC Press, Boca Raton, FL, USA.

Sivertsvik, M. (2007). The optimized modified atmosphere for packaging of pre-rigor filleted farmed cod (*Gadus morhua*) is 63 ml/100 ml oxygen and 37 ml/100 ml carbon dioxide, LWT - Food Science and Technology 40(3):430-438.

Stoffers, N.H., Dekker, M., Linssen, J.P.H., Störmer, A., Franz, R. and Van Boekel, M.A.J.S. (2005). Modelling of simultaneous two-sided migration into water and olive oil from nylon food packaging, Eur. Food Res. Technol 220:156-162.

Toepfl, S., Mathys, A., Heinz V. and Knorr D. (2006). Potential of High Hydrostatic Pressure and Pulsed Electric Fields for Energy Efficient and Environmentally Friendly Food Processing, Food Reviews International 22(4):405-423.

8. Aspects of hygienic design

Huub L.M. Lelieveld
President of the European Hygienic Equipment Design Group (EHEDG) 1989-2002
and Honorary Life Member of the Executive Committee

8.1. Introduction

The first prerequisite for designing foods is that the resulting products will be safe. This implies that food plants (equipment, walls and ceilings of buildings, indoor and outdoor surroundings) should be designed in such a way that this can be achieved in practice. In addition, the plant should be built in carefully selected areas, to avoid an undesirable influence (microbiologically and toxicologically) of the environment on the safety and quality of the food. Problems caused by excessive microbial contamination of foods tend to be expensive, particularly if they result in public recalls. As a result of the application of increasingly mild preservation technologies, the use of modified atmospheres and chilled storage, processed foods become more sensitive to microbial (re)contamination, requiring greater control of the manufacturing process.

The hygienic design and location of food factories and equipment play an important role in controlling the microbiological and chemical safety and quality of the products made. This is officially recognized in legislation, such as the various regulations of the European Commission, concerning the hygiene of foodstuffs (European Parliament and Council of the European Communities, 2004a, b, 2006). The Codex Alimentarius (Codex Alimentarius Commission, 1996) also addresses hygienic design in relation to food safety.

A hygienic factory should prevent products from:
- having too high microbial counts;
- containing toxins of microbial origin;
- containing residues of chemicals used for cleaning and disinfection;
- being contaminated with other non-food substances such as lubricants, coolants and antimicrobial barrier fluids;
- containing foreign bodies, such as pieces of metal, plastic, packing material and insects (or other vermin) or parts thereof.

A factory which has been designed with hygienic requirements in mind, with process lines correctly built from hygienic equipment *and* properly operated and maintained, will produce food products that are safe and of excellent microbiological quality. This chapter deals with the influence of the design of a plant on the quality and safety of the product and discusses hygienic requirements. It is based upon a much more detailed chapter by the present author in an earlier book (Lelieveld, 2000). A preliminary remark is that there are many types of food and microbial vulnerability therefore varies widely. The only way to decide which measures are needed is to carefully estimate the risk for each product and every single process step. In other words, hygienic design can only be done properly with sufficient knowledge of food science and technology, or in yet other words: hygienic design should be an integral part of product and process design.

8.2. Possible threats

8.2.1. High microbial counts

Too high microbial counts may be the result of high counts in the raw materials or the survival of micro-organisms due to a failure of the process control or equipment; too high counts may also be caused by recontamination of product. A process may fail because of too short treatment times. Inefficient mixing may cause an inhomogeneous temperature distribution and the measured temperature may be different from – most importantly - the temperature of the coldest part of the product. Too low temperatures in thermally processed food may also result from the build-up of food or food-derived materials on heat exchanging surfaces.

Incorrect design of the factory may result in mixing raw with decontaminated product or the transfer by other means (e.g. air, personnel or equipment) of micro-organisms from raw to finished product ('cross contamination'). Equipment that is correctly cleaned and sanitized before use may be recontaminated by the product as a result of an incorrect start-up procedure; the thus contaminated equipment may subsequently recontaminate passing product. Furthermore, individual components of process lines, if of incorrect design, may cause recontamination. Incorrectly designed process lines, even if properly cleaned (e.g. manually with disassembly where needed), may lead to contamination of product and thus to high microbial counts because micro-organisms may multiply to high numbers in areas where product is stagnant ('dead' areas) and may release large numbers to passing product.

The process equipment itself may still be contaminated after cleaning and sanitation as a result of not being cleanable (due to the presence of non-cleanable features such as screw threads and crevices), or not being given the required treatment (because it is not resistant to the heat or biocidal chemicals), or just not being properly cleaned because the cleaning procedure is incorrect or the control system does not function or has not been set up properly.

In an open plant, where product is exposed to the environment, microbial contamination may come from the air, from dust, from condensate falling down from e.g. chilled pipelines, from splashes of other liquids, insects (or sometimes even birds), and of course directly from people touching the product. The risk of contamination is inversely proportional to the cleanliness of the plant and hence to its cleanability, which in turn depends on the site where the factory is located and the design of the building (Lelieveld and Holah, 2011), and the design and installation of all that is inside the building (Lelieveld *et al.*, 2003, 2005).

Where microbial growth in the product is controlled by its composition, failing sensors or incorrectly calibrated measurement equipment may result in deviations in, e.g. pH, water activity and concentration of preserving agents. Finally, the primary packaging material may be unacceptably contaminated, because the quality as delivered is not good enough or - in case of aseptic packaging - because of inadequate decontamination or recontamination of the material.

8.2.2. Toxins of microbial origin

Micro-organisms may produce toxins that survive heat, such as *Staphylococcus aureus* enterotoxin; and many mycotoxins (produced by moulds), such as aflatoxins, ochratoxins, trichotecens, fumonisins and zearaleons, are also heat-stable. In other words, cooking, pasteurization and sterilization do not destroy such toxins. Therefore, if a product - at any stage in the product chain - has acquired too high a count of toxigenic micro-organisms, that product should never be upgraded by destruction of the micro-organisms, but should be destroyed to avoid it from being used.

8.2.3. Chemical residues

Chemicals used for cleaning and disinfection of equipment or packing material should never be left behind to contaminate the product. The concentration of any residues should not exceed the legally accepted level. For instance, in many countries the product may not contain more than 0.5 mg hydrogen peroxide per kg product. Preferably, residues should be non toxic. Intrinsically toxic products such as tetraborate and chloroisocyanuric acids should be avoided in food processing plants.

Construction materials, i.e. metals, polymers, elastomers, but also auxiliary substances such as lubricants, antimicrobial barrier liquids, coolants and signal transfer fluids, may contain toxic substances that may migrate to the product, depending on the product characteristics (such as pH and chloride contents). Therefore, materials used for the construction of components that may come in contact with food, should be selected very carefully. Under the conditions of use these materials should not be able to make a food product toxic (European Parliament and Council of the European Communities, 2004c).

8.2.4. Foreign bodies

Foreign bodies may originate from raw materials (bones, insects) and packing material (wood, cardboard, metal, glass, polymers), but also from the building (glass, wood, chalk, paint flakes, screws, nails), equipment (bolts and nuts, security rings, pieces of gaskets, metal and polymer shavings), from the environment (insects, rodent and bird droppings, dust), and from people (jewellery, pens, pencils, hair, hairpins, buttons, nails) and peoples' activities (tools, bristles from brushes, pieces of cloth). The design of the plant should prevent foreign bodies from contaminating the product at all stages of the manufacturing process. The correct choice of construction materials, measures to prevent the presence of insects, rodents and birds and facilities for personnel to safely store their belongings and their tools, will greatly reduce the risk of foreign body incidents.

8.3. Factory design

The production plant must be built in such a way that the finished product cannot be recontaminated by raw materials (including packaging materials), or any intermediate product. Thus, the flow of material should be from the raw materials reception to the - separate - finished product exit. Similarly, it should not be possible for personnel working with raw materials to move to or pass through upstream areas, where products may become contaminated. Stores should be located and provided with entrances and exits in such a way that no less-clean areas need to be crossed to obtain clean materials. Functions not directly involved in the manufacturing process, such as administrative offices and canteens, should be separated from the product areas to prevent staff from entering these areas. Preferably, offices and canteens should be in a separate building. If not, they should be separated from the product areas by a corridor.

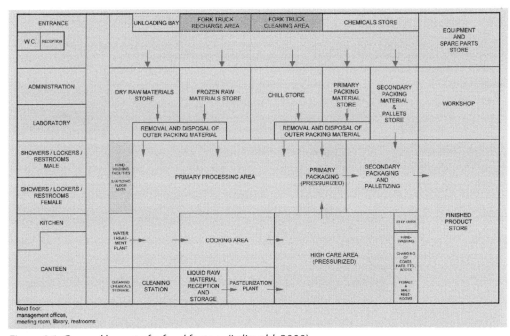

Figure 8.1. General layout of a food factory (Lelieveld, 2000).

Figure 8.1 gives a generalized example of the layout of a food factory, assuming that there is a process step following the steps where the relevant micro-organisms are inactivated, and not followed by another process step, which would provide the required inactivation. Arrows indicate the (compulsory) direction of the flow of materials. A hazard and risk analysis should be used to determine which areas are needed, taking into account product vulnerability and shelf-life requirements.

The position of any air inlets and outlets should be carefully chosen to avoid cross contamination of products. The air should move from the cleanest side or sides (where the finished products are) to the least (microbial) clean sides (where the raw materials are). It should be ensured that this is done correctly, which means, for instance, that the location of doors and windows should be taken into account. In some cases doors must be interlocked to ensure that only one of them can be open at a time. To ensure that safety exits are not used as normal doors, they may have to be provided with alarms or locked with a special key, which is kept behind a glass panel to be broken in case of emergencies.

All surfaces must be cleanable, which must be considered when selecting construction materials, and must be able to withstand the chosen cleaning methods, temperatures and chemicals. Cleaning methods required are usually determined by the type of contamination (incl. product residues) to be removed. Walls may have to be resistant to high-pressure jet cleaning, and floors to mechanical cleaning with brushes; all surfaces may have to be resistant to chlorides and hypochlorite, hot alkali and hot acid solutions. Decisions on cleaning methods should be made before, or in conjunction with, decisions made about construction materials.

If materials are stored for relatively long periods of time (several days or more), the growth of moulds may become a serious problem if the humidity in the storage area is higher than 55%. It is therefore important to keep the humidity below that value. Depending on local conditions, ventilation may be sufficient. If, however, microbial susceptible materials, which include cardboard, used abundantly as packing material, are stored for more than a few days and the humidity in the environment is high, it may be necessary to install air conditioning or dehumidification equipment. In some cases it may suffice to use conditioned air from the production area. Alternatively or in addition, the temperature in the relevant storage area may be reduced.

Production areas may be divided into areas where the product is exposed, the so-called 'open plant', and areas where product is enclosed, called 'closed plant'. In open plants there is self-evidently a higher risk of contamination of the product from the environment. During primary packaging the product is exposed and thus the area where the product is packed should be treated as open plant, even if the product is delivered from a closed process line.

Many requirements apply to both open and closed plants. Production areas should be designed such that vehicles coming from outside the factory cannot enter. Instead, materials should be transferred to dedicated indoor vehicles. Insect-proof screens with a sufficiently fine mesh should be used where open windows are used. There must be no niches or recesses where insects can hide. Equipment must be sealed to the floor or wall or enough clearance must be left for inspection and cleaning. Surfaces should be non absorbent unless only dry products are produced and the production area will always be dry.

Whether the product is exposed or not, the risk of contamination of the product from the environment will always be proportional to the concentration of micro-organisms in that environment. It is therefore important that multiplication of micro-organisms in the production area is prevented. As microbial growth is possible only in the presence of enough moisture, such growth can be prevented by ensuring a sufficiently low humidity. As moulds will grow at water activities down to approx. 0.6, keeping the relative humidity at or below 50% is highly recommended. A humidity of 50% is also comfortable for human beings.

Water vapour may condense on cold surfaces, causing excessively high water activity locally. Therefore, water vapour should be properly vented near to where it is generated and the installation of hoods may be needed. This may be the case where product is cooked, where vegetables are blanched, or where equipment or utensils are cleaned and sanitized using hot solutions or steam. The inside of such a hood should be regularly inspected for cleanliness as there the humidity will probably be high. It is likely that at least a part of the surface is at a temperature suitable for microbial growth, for at least part of the time. Pools of water are ideal places for micro-organisms to grow, in particular in food factories where the likelihood that sufficient nutrients are present is very high. The presence of pools of water and wet floors will also make it difficult to keep the humidity low.

Where dust is generated (handling of dry powders, spices, etc.), care must be taken that the dust is extracted at that spot. For emptying bags, big or small, dust-free emptying systems are available, combined with dust-free packing material compactors. Depending on local conditions, in particular the location of the factory, the microbiological product requirements and the preservation processes used, it may be necessary to install air filters. Normal air carries micro-organisms. In many environments the concentration of micro-organisms in the air ranges from 100 to 1000 per cubic metre. In agricultural areas the concentration may be much higher and vary with the season. In coastal areas the concentration is often low.

Airborne micro-organisms may be single spores, e.g. conidiospores of moulds like *Penicillium* and *Aspergillus*. Bacterial spores, however, often occur as clusters, surrounded by a mixture of anorganic and organic material, remnants of the substrate in which they have developed. High concentrations of vegetative bacteria are usually found close to their origin only, e.g. near waste water treatment plants, although some vegetative bacteria, such as some micrococci are tough enough to survive in air for a long time. The origin, of course, can also be sneezing, coughing and talking human beings who suffer from infectious diseases. In many instances, airborne micro-organisms are associated with dust particles. Air can be sterilized within less than a second by being heated to a temperature of 350 °C (Elsworth, 1972). At such a temperature, organic materials, including all known micro-organisms, independent of their dimensions, disintegrate. Incinerators are therefore probably more reliable than filters, provided that their design guarantees that all air is heated to the required temperature. Therefore, an incinerator must be equipped with a reliable temperature control system. Although this method does not require much energy because of the low heat capacity of air, the method is more expensive than filtration for most applications.

Filtration is based on various mechanisms of particle retention. Depth filters usually rely on the entrapment of particles in a labyrinth of fibres and electrostatic attraction. Membrane filters rely on pore sizes. Membrane filters have a much higher resistance than depth filters and are therefore not normally used in the food industry. Industrial depth filters are made from corrugated sheets of compact fibrous material, which is mounted in cartridges using a glue or resin. The cartridges in turn fit in the filter houses, which are mounted in the air ducts (air for clean rooms) or pipelines (process air). As a leak may virtually annihilate the effect of the filter (if 1% of air is passing between casing and cartridge, the reduction in concentration of micro-organisms will be less than 100-fold), the cartridges must be carefully installed. For critical applications it may be advisable to use two filter assemblies in series and/or to have the filter assemblies tested before installation and use (Vinseon, 1992).

8.3.1. Personal hygiene

Hand-washing facilities should be installed near the entrance to the production area so that everybody entering the area passes these facilities. The taps should be operated without using hands, e.g. by foot or elbow. Alternatively, electronic detection systems may be used. The facilities should include a supply of single-use (e.g. paper) towels and cleanable waste bins. Rooms for changing clothing and footwear should also have adequate facilities for the cleaning of hands and should be equipped with lockable facilities for personal belongings. Rest rooms (washrooms, toilets, flush lavatories, WCs) should be easy to clean and be provided with adequate means for cleaning, disinfection and drying of hands. Preferably, the operation of taps, dispensers, dryers and doors should be automatic to minimize contamination risk. In any area where direct or indirect contact with food is possible, nobody should be allowed access without first taking a number of precautions aimed at avoiding product contamination risks.

8.3.2. Packaging

During packaging, the product and often the product side of the packing material are exposed to the environment. The time of exposure, the concentration of micro-organisms in the air and the sedimentation rate of the micro-organisms will determine the degree of infection of the product. Even if packing is done aseptically, the rate of contamination of the packed product is influenced by the concentration of micro-organisms in the surroundings of the packing machine. Depending on the acceptable contamination rate, the concentration of micro-organisms in the area of the aseptic packing operation may have to be reduced.

8.3.3. High-care areas

In addition to the measures which apply to all processing areas, extra measures may be required to keep contamination of the exposed product under control. The measures taken depend on the product and the intended storage, distribution and shelf life. A proper risk assessment should be carried out to establish the measures needed to meet the targets set. Measures may include:

- A change area with a step over, which allows for the exchange of footwear without contaminating the clean side with footwear from the other side.
- Supply of filtered air.
- Maintenance of a higher pressure to reduce ingress of contaminated air.
- The use of gloves. When using gloves to protect the product, no unclean surfaces should be touched. The gloves must be replaced frequently. As they promote the growth of micro-organisms on the skin, care must be taken that these micro-organisms cannot contaminate the food product. Therefore, changing gloves should be done well away from the product (or primary packing material).
- Mouth masks may be required. They are usually effective, provided they are replaced often enough.
- If the multiplication of micro-organisms must be prevented during the processing in the high-care area, which is often the case with products of animal origin, the high care area should be chilled, typically to a temperature below 12 °C (European Parliament and Council of the European Communities, 2004b). The actual temperature will depend on the possible rate of microbial growth in the product and the time the product resides in the area. Unless the residence time in the packaging areas is very short, these areas must also be cooled. Adequate clothing, which protects against the cold, must be provided. The clothing must nevertheless comply with the hygiene requirements listed above.

8.4. Equipment design

The function of equipment is to convert raw materials into wholesome food products. If equipment is not designed correctly or is assembled in the wrong way, the wholesomeness of the food may be impaired. Equipment may be the source of microbiological problems for a number of reasons. It may be that the equipment:

- cannot be cleaned and decontaminated, and allows microbial growth causing contamination of the product in contact with the equipment;
- should decontaminate (pasteurize or sterilize) the product but fails to do so;

- although cleanable, is such that part of the product resides in certain areas for too long a time, giving micro-organisms time to multiply;
- does not give the required protection against recontamination from the environment.

The choice of equipment used to build a process line should be based on risk assessment, which takes into account the microbial susceptibility of the product, the stage in the process and the cleaning frequency. For handling raw materials, straight from the field, the requirements will be different from those for washed or cooked product. The aim should be that none of the process steps decreases the microbiological quality of the product. This also requires measures to prevent cross contamination. Having selected the individual components, it is extremely important to ensure that in the design of the whole process line cleanability, sanitizability (disinfection, pasteurization, sterilization) and for closed process lines for aseptic processes, bacteria-tightness are retained. Measurement and control systems that are used for critical process steps must be designed in such a way that measurements are correct at all times. The critical process steps include those which necessarily precede the actual production, i.e. cleaning and decontamination of the process line. In case of any system failure, an alarm must be given and in most cases the process must be stopped automatically.

For process lines not intended for sterilized products, the components must at least be pasteurizable. Downstream from the heating section of any sterilization plant requirements may differ. If the sterilized product needs to have a long shelf life at ambient temperature, all components must of course be sterilizable. In other cases, e.g. if the sterilized products are meant to have a limited chilled shelf life, it may be sufficient for the equipment to be pasteurizable. Long shelf life products, whether made microbially stable by sterilization (low-acid products of high water activity) or by pasteurization (acid products or products with a sufficiently low water activity) must be protected against recontamination after the heat treatment and thus require equipment which is bacteria-tight.

Materials intended for contact with food must be corrosion resistant, non-toxic, mechanically stable, have a surface finish that is not adversely affected under conditions of use and be smooth enough to be easily cleanable (Hauser *et al.*, 2004). No toxic construction materials should be used and the relevant legislation must be respected. The traditional stainless steel materials, such as types AISI-304, AISI-316 and AISI-316L are fully acceptable for most applications (i.e. under the conditions that these materials are corrosion-resistant). Care must be taken when polymer and elastomer materials are used as they may contain leachable toxic components. The same applies to the use of adhesives, lubricants and signal transfer liquids. Greases and lubrication oils should comply with §178.3570 of the Food Drug Cosmetics Law Reports (Food and Drug Administration, latest update). The supplier must present evidence that the material is safe for use in contact with food.

To retain the desired smoothness of surface, the equipment must be resistant to the product under the process conditions and withstand the cleaning procedure (concentration of chemicals at temperature of use for the times needed, e.g. 250 mg/kg sodium hypochlorite at 40 °C for 20 min) during the intended life time. Corrosion can be minimized by sticking to the concentrations, times and temperatures specified, and by complete removal of residues (or neutralization if any is left at non-product contact surfaces of equipment). Many food products contain chloride and have a pH between 3 and 5, a very corrosive combination. Often the rate of corrosion is determined by the product because, in addition, contact with the product is much longer than the contact with cleaning and sanitizing agents.

8.4.1. Decontamination of equipment

Micro-organisms can be very well protected by residues from destruction by heat. Sterilization with steam is dramatically impaired by the presence of soiled crevices (Lelieveld, 1996). For instance, treatment with saturated steam at a temperature of 120 °C for 30 min. will inactivate all viable *Bacillus subtilis* B1-12 spores on the surface of clean equipment, the D_{120}-value of these spores being approx. 14 s (Put and Aalbersberg, 1967). In residues, consisting of normal food constituents, i.e. protein, sugar, fat or a mixture of fat and protein, the D_{120} values found were 13 to 28 min., at water activities of 0.5 to 0.6. Furthermore, the reduction in the number of viable spores in the protein/fat mixture dropped to a factor of 1000 in a crevice at a distance of 15 mm from the surface. The decrease in degree of inactivation cannot be attributed solely to the slightly lower temperature in the crevice (approx. 117 °C, for more than 10 min.). More importantly, the water activity in the residue remains low despite the saturated steam passing the entrance of the crevice. D-values increase dramatically with a decrease in water activity (Collier and Townsend, 1956). This applies to all types of micro-organisms, including aerobic and anaerobic bacterial spores (Collier and Townsend, 1956), vegetative bacteria such as staphylococci (Verrips and van Rhee, 1981), as well as moulds, e.g. *Aspergillus* species (Doyle and Marth, 1975). Thus, crevices may have a very dramatic effect on the sterilization of food processing equipment and hence on product safety. Clearly, it is essential that equipment is clean before decontamination.

Open plant equipment is usually decontaminated using microbiocidal chemicals at near ambient temperatures, such as sodium hypochlorite, typically in a concentration of 50 - 200 mg/kg (total elemental Cl) for 1 - 10 min., and quaternary ammonium compounds, 100 - 200 mg/kg. To save energy, closed plants are often disinfected chemically as well. Closed process lines are usually decontaminated by thermal methods. If bacterial spores are not relevant (e.g. the line is used for pasteurized products), hot water is used. If bacterial spores are relevant, decontamination is carried out either with steam or hot water (under pressure).

The European Hygienic Equipment Design Group (EHEDG) has defined standard methods to test the pasteurizability (Venema-Keur *et al.*, 1993) and sterilizability

(Timperley *et al.*, 1993a) of equipment. These methods should be used if there is any doubt as to whether a piece of equipment, which must be free from (relevant) micro-organisms before use, can be pasteurized or sterilized. In these methods, the equipment to be tested is contaminated with a test organism (ascospores of *Neosartoria fischeri* for pasteurization and *Bacillus subtilis* spores for sterilization). Then the equipment is pasteurized (with water of 90 °C for 30 min.) or sterilized (with saturated steam of 120 °C for 30 min.). Subsequently the equipment is filled, aseptically, with nutrient medium to check for growth of the test organisms. The bacteria-tightness of equipment can be tested by a standardized method (Timperley *et al.*, 1993b), which is often performed in conjunction with the sterilization test described earlier. In this test, equipment is sterilized and filled with a sterile liquid culture medium. The exterior of all suspected areas of the equipment to be tested are wetted with a culture of *Serratia marcescens*, a very motile bacterium. This is repeated several times, while any movable component is actuated. If the medium in the equipment remains sterile, it is apparently bacteria-tight.

If a product allows the growth of micro-organisms, and micro-organisms are present, multiplication will almost always take place, even if all possible measures appear to have been taken to prevent multiplication. The geometry of equipment influences residence-time distribution and there are usually places in process lines, even if correctly designed, where some product resides longer than desirable. Even if dead areas have been designed out, there is always a certain amount of product attached to equipment surfaces, even at high-liquid velocities. Trapped micro-organisms may reside on such surfaces long enough to multiply. With the increase in the number of micro-organisms, the numbers washed away with the product passing increase as well.

8.4.2. Biofilms

Biofilms are collections of micro-organisms, extracellular polymeric products, and organic matter located at interfaces. Depending on the properties of the surface and the types of micro-organisms present, with time biofilms may develop that are very hard to remove. Biofilm development may take place on any type of surface and is difficult to prevent if the conditions sustain the multiplication of micro-organisms. Many micro-organisms, including many pathogens (*Listeria monocytogenes*, *Salmonella typhimurium*, *Yersinia enterocolitica*, *Klebsiella pneumoniae*, and *Legionella pneumophia*) form biofilms, even under hostile conditions, such as the presence of disinfectants. Actually, adverse conditions seem to stimulate micro-organisms to grow in biofilms (Van der Wende *et al.*, 1989; Van der Wende and Characklis, 1990). Thermophilic bacteria (such as *Streptococcus thermophilus*) can form biofilms in the cooling section of milk pasteurizers, sometimes within five hours, resulting in massive infection of the pasteurized product (up to 10^6 cells per ml) (Driessen and Bouman, 1979; Langeveld *et al.*, 1995). Biofilms shed micro-organisms on to the passing product. On metal surfaces (including stainless steel), biofilms may also enhance corrosion, which, in plate heat-exchangers, may result in microscopic holes. Such pinholes are known to allow the passage of micro-organisms present in the cooling medium and thus may cause infection of the product.

Like other causes of fouling, biofilms will affect heat transfer in heat exchangers. On temperature probes, biofilms may seriously affect the heat transfer and thereby the accuracy of the measurement. On conveyer belts and on the surfaces of blanching equipment, biofilms may infect cooked or washed products, which are assumed to have been made pathogen-free by the temperature treatment received. Biofilms may be much harder to remove than ordinary soil. If the cleaning procedure is not capable of completely removing biofilms which may have developed, decontamination of the surface by either heat of chemicals may fail as biofilms dramatically increase the resistance of the embedded micro-organisms. It is therefore imperative that product contact surfaces are well cleaned before disinfection. Wirtanen (1995) gives an excellent overview of fouling in relation to equipment for the processing of food.

In attempting to prevent the growth of micro-organisms in process lines by cooling, one should be aware that at all places where there is activity, energy is converted into heat, causing rises in temperature. The consequence is that in chilled rooms (of any size) and in pipelines with cooled products there are spots of higher temperature, where micro-organisms multiply and subsequently infect passing product. Typical examples are mechanical seals, where the friction between the static and the rotating part dissipates a considerable amount of energy, and shafts of electro motors, which conduct the heat of the electric drive to the device in contact with the product. These effects can be significantly reduced by using barriers made from materials with a low thermal conductivity (e.g. polymers or elastomer flanges between the motor and the driven piece of equipment). Cooling may also be applied (e.g. in mechanical seals).

The subject of the growth of micro-organisms in process lines has been given considerable attention as it may cause problems whose origin is often not recognized. The potential importance can be illustrated by calculating the infection of a product as a result of the presence of stagnating product. If a cell of *E. coli* is trapped in a dead space filled with 5 ml of a slightly viscous low acid food product at a temperature of approx. 25 °C, it may take less than 24 h for the number of *E. coli* cells to increase to a concentration of 0.2×10^9 per ml; assuming a doubling time of 40 min. If every hour just one ml is washed out from the dead space by the passing product, by the end of the first day of production, the product is infected with 200 million *E. coli* cells every hour. If the production capacity of the line is 5 million ml per h, the average *E. coli* contamination will be 200/5=40 per ml. Many traditional process lines have much larger (often very contaminated) dead spaces and temperatures and growth rates can be higher.

8.4.3. Packaging

If a product has been produced and pasteurized or sterilized correctly, it must obviously be ensured that the next step, namely packaging, does not adversely affect the condition of that product. Hence, care must be taken that no unacceptable recontamination takes place during the transfer of the product to the packaging equipment and during the packaging operation itself. If the process line and the packing machine are cleaned and decontaminated independently, which is usually the case, it must be ensured that making the connection between the two does not impair the hygienic condition achieved. In the case of aseptic packing, making the connection must also be done in an aseptic way. This requires the thorough decontamination of the connection before starting the flow of product. Equally, during cleaning and decontamination of the packing machine the process line or aseptic storage tank must be protected against contamination by the cleaning chemicals and *vice versa*.

The risk of contamination of a surface with micro-organisms will always be proportional to the concentration of micro-organisms in the immediate environment and the time of exposure. Therefore, to reduce the risk of infection during packaging, it may help to reduce the concentration of micro-organisms in the environment and to keep the time of exposure to the environment of the product side of the packing material and the product to a minimum. The sedimentation velocity of micro-organisms present in the air is in the order of v_s=0.003 m per s (L.H. Kerkhof, R. de Weijer and H.L.M. Lelieveld, personal communication). A simple calculation shows that the infection of a product by micro-organisms in the air can indeed be significant: if the concentration in the air is x_{air}=1000 per m^3, the time of exposure is t=5 s and the exposed surface area is A=0.01 m^2 (i.e. 100 × 100 mm), the number of micro-organisms settling on that surface will be n = v_s × x_{air} × t x A = 0.15. In other words, under the conditions assumed in this example 15% of the surfaces exposed will be infected with one micro-organism. This may or may not be acceptable in case of hygienic packing; it will certainly be unacceptable for aseptic packaging. The risk of contamination may be reduced by simple measures, such as the installation of a tunnel over the area where the packing material and the products are exposed. If this is not effective enough, sterile air may be supplied to the tunnel.

If the shelf life of the product to be packed is determined by its intrinsic stability, i.e. the product is dry or preserved by acidity, water activity (a_w), or chemical preservatives, if the required shelf life is short or if the desired shelf life is obtained by cold storage, freezing or in-pack heat treatments, it is quite acceptable for the product to contain a number of micro-organisms when ready to pack. Unless an adequate in-pack heat treatment is to be given, such products should not be contaminated with pathogenic micro-organisms. In any case, the concentration of micro-organisms should not exceed the legal limits or limits which are determined by the organoleptic quality at the ultimate consumption date.

Aseptic packing is aimed at products with a long shelf life at ambient temperature. The acceptable presence of relevant micro-organisms in such products is very low (e.g. less than one per 10,000 packs). As to which micro-organisms are relevant, that depends on the type of product. For acid products, such as many fermented dairy products and fruit juices, moulds and yeasts are usually relevant, but bacteria (in particular spore-forming bacteria) are often irrelevant. For low-acid products with a high water activity, virtually all micro-organisms may be relevant. To achieve the required low risk of contamination, apart from controlling the microbiological quality of the air and packing material, the time of exposure of product and packing material and decontamination of the material by a suitable means, attention will have to be paid to complete sterilization (or in the case of acid or low water activity products: complete pasteurization) of the packing machine. Furthermore, the product supply line, including the transfer pump, must be bacteria-tight. To avoid contamination after packaging, the pack filled must remain bacteria-tight and hence the integrity of the filled and sealed pack must be very good.

To avoid resources being spent on the wrong measures, careful determination of which micro-organisms *are* relevant is essential. Subsequently, the acceptable risk of contamination with those micro-organisms must be established. Having done so, a risk analysis should be carried out, which requires a careful inventory of all potential sources of relevant micro-organisms. This approach is well reviewed by Cerf and Brissenden (1981).

8.5. Design of process lines

It must be ensured that all product experiences the correct temperature for the time intended. The temperature probe must be mounted in a position that ensures that it measures the temperature of the product passing and not, for example, of stagnating product. If in an awkward position, the probe might measure the average of the temperatures of a nearby steam pipe and the product in the line, and thus report a higher temperature than the product actually has. If this or a similar situation is indeed possible, the section of pipeline with the probe must be thermally insulated.

As the residence time is proportional to the volume of the holding section, the potential of the product to cause fouling must be seriously considered, as fouling may result in a thick layer of product on the wall, reducing volume and hence reducing residence time. It should be determined when fouling occurs. The process must be stopped before the fouling reaches unacceptable levels.

After leaving the holding section, the product that still has to be pasteurized or sterilized is used to cool pasteurized or sterilized product in the regeneration (heat recovery) section of the heat-exchanger. Further cooling is done using a regular cooling medium (e.g. water with or without additions). In both cooling sections the treated product must be adequately separated from the raw product or the coolant, as both are likely to contain micro-organisms. To reduce the risk of contamination, there must always be two gaskets separating the treated product from the cooling liquid and it must be impossible to build up a pressure in the space between the gaskets. Hence, there must be grooves between the gaskets large enough to ensure that any leaking product or coolant flows freely to the outside.

To minimize the risk of recontamination, the raw and treated product may be separated from each other by a separation loop. Such a loop, if well maintained, does not dramatically increase the cost of the heat-exchanger and decreases the heat recovery by no more than five percent.

If a pasteurizer with a heat-recovery section and flow diversion is connected to either a filling machine or a pasteurized product storage tank, it is common for the packing machine or storage tank to be decontaminated separately (often having been delivered by a different supplier from the one delivering the process plant). To ensure that the process line can be decontaminated (in this case pasteurized) without contaminating the packing machine or storage tank, a recirculation valve is needed between the process line and the machine or tank. If the pasteurized product is intended for a long shelf life at ambient temperature, filling must be aseptic. In such a case, the components downstream from the heat-exchanger must be bacteria-tight. Details on microbially-safe continuous pasteurization of liquids are given by Lelieveld *et al.* (1992).

If steam injection is used to heat the product, care must be taken that the injector is of hygienic design, both inside and out, as both sides will be in contact with the product. It must be made impossible for residues of cleaning chemicals to become trapped in the injector.

8.5.1. Refrigeration of food products

The refrigeration of food products, applied to prevent microbiological problems, has been considered intrinsically safe for a long time as low temperatures stop the multiplication of micro-organisms. As a consequence, refrigeration systems have often not been designed to be easily cleanable. Many micro-organisms, including pathogens, are now known to be able to grow at refrigeration temperatures. They do grow when given suitable conditions, such as those created in refrigeration systems. Obvious examples are condensate trays placed underneath cooling aggregates, gutters and drain traps. There are, however, also hidden areas with suitable conditions, such as crevices, sharp corners and service conduits. Water collected in such places may reside there for long periods of time, so that micro-organisms, even if starting from a single cell, may grow into many millions. The air currents, essential for heat transfer, efficiently distribute the micro-organisms over the exposed product.

When there is refrigeration, there are temperature gradients. Self-evidently, there is a gradient between the environment and the refrigerated area. There are always connections between the warmer environment and some of them may actually be with relatively hot objects, such as motor shafts of e.g. fans, freezer barrels and mixers. Some parts of the equipment may produce heat, such as lighting, bearings and mechanical seals. The consequence is that there will be areas that are supposed to be refrigerated, which are at a significantly higher temperature than assumed. In such areas, in the presence of moisture or product, microbial growth will be faster than may have been anticipated.

If equipment (including freezer tunnels, conveyors, fans, refrigerated enclosures, freezer barrels, lighting armatures, doors, etc.) is not designed hygienically, the amount of hidden liquid and product residues still present after cleaning can be large. Chemical disinfection is not effective in the presence of dirt and the effect of heat treatments is greatly diminished. As it may take a long time before the refrigeration temperatures are reached again, micro-organisms will continue to multiply and may reach concentrations of 10^9 per g, waiting to be mixed with or spread over the product. Therefore, refrigeration equipment should also comply with hygienic requirements as discussed in this chapter. There are good reasons why 'refrigeration' is mentioned explicitly in the relevant EU directives (European Parliament and Council of the European Communities, 2004a).

8.6. Legislation

The European legislation with respect to food safety (European Parliament and Council of the European Communities, 2004a, b, c) is firm and clear with respect to the objectives. The same applies to legislation in the USA (e.g. Food and Drug Administration, 1989, 1997). The legislation - correctly - does not give details on *how* to comply but sets down what standards should be achieved. The initiatives and responsibilities for achieving these standards are left with the manufacturers of food as well as with the suppliers of the machinery used to produce the food. In 1989 the European Commission mandated the European federation of standardization organizations (CEN) in Brussels to produce standards on the hygienic aspects of food processing machinery. Since then, much work has been done by CEN's Technical Committee 153 and its many working groups. Many standards have been prepared. The EHEDG (Anonymous, 1992) has published more than 40 bulletins on how to comply with hygienic requirements for the safe production of food and how to verify whether equipment complies with the requirements for hygienic or aseptic operation of food plants; several independent institutes in Europe will test equipment according to the EHEDG recommendations (Cerf *et al.*, 1995; see also www.ehedg.org), The EHEDG has been working together with a similar organisation in the United States, the so-called 3-A organisation, since 1992, with the objective of harmonizing principles and requirements between Europe and North America. The International Dairy Federation (IDF) is also active in the area of the hygienic design of dairy plants (see also Cerf and Brissenden, 1981).

8.7. Concluding remarks

To avoid unnecessary hygiene problems, it is vital that when contracting to build a food factory an appropriate site is selected and hygiene requirements are clearly made a part of the contract. Similarly, when ordering equipment and other materials for a food production plant, it is essential that the user specifies his requirements clearly. It is also very important that the user checks whether contractors and suppliers understand what is being asked of them (Vinseon, 1992), because repair afterwards may be extremely costly. It is important to make sure that the plant or process line will deliver safe products. Validation of the cleanability and sanitizability and – where relevant – imperviousness to micro-organisms is therefore an essential step (Carleton and Agalloco, 1986). Food should be safe not only immediately after commissioning of the plant; the plant or installations must continue to comply with hygienic requirements throughout their life cycle. Therefore, an appropriate scheme for preventive maintenance should be devised. It is obviously also necessary to make hygienic design an integral part of food quality management systems (Lelieveld, 1994; Luning *et al.*, 2002).

Although in the past 20 years or so food processing equipment has become increasingly available and factories have been built with clearly improved hygienic performance, there are still many cases where there is still a long way to go with respect to hygiene. For a large part this is due to lack of awareness. Many individuals with responsibilities for the handling and processing of food have little understanding of hygiene. The result is hundreds of thousands food poisoning incidents per year. This applies to the entire food chain, from the farmers who actually produce the raw materials to the retailers and caterers who deliver the final product to the consumer and, indeed, the consumer, who often does his or her own final processing. It is hoped therefore that in the near future in cooperation with all those involved in the food chain, training and education in food hygiene will be improved, resulting in increased awareness. This will contribute greatly to safer food.

In addition, with the developments in electronics, sensors will become available that are able to measure minute changes in physical parameters, including temperature, pressure, light, sound and ultrasound. These developments may be used to make increasingly safer food. Appropriate research and development will lead to devices, which will give a warning when fouling starts or approaches the acceptable limit, when cleaning of critical surfaces is completed, when filters leak or become too resistant, when deviations from the expected flow and direction of air occur and when the concentration of particles in the air is increasing. Gradually, such devices will tell the plant operator when to initiate which actions to ensure maximum product safety and quality.

References

Anonymous (1992). European Hygienic Equipment Design Group. Trends in Food Science and Technology 3(11): 277.

Carleton F.J. and Agalloco J.P. (eds.) (1986). Validation of aseptic pharmaceutical process. Marcel Dekker, New York.

Cerf, O. and Brissenden, C.H. (1981). Aseptic packaging. IDF Document 133. International Dairy Federation, Brussels, pp. 93-104.

Cerf, O., Axis, J., Maingonnat, J.-F., Traegarth, C., Hodge, C., Kirby, R., Holah, J., Kastelein, J., Poulsen, O., Roenner, U. and Mattila-Sandholm, T. (1995). Experimental rigs are available for the EHEDG test methods. Trends in Food Science and Technology 6(4):132-134.

Codex Alimentarius Commission (1996). Draft Revised Recommended International Code of Practice. General Principles of Food Hygiene. Alinorm 97/13. Food and Agricultural Organization of the United Nations World Health Organization, Rome.

Collier, C.P. and Townsend, C.T. (1956). The resistance of bacterial spores to superheated steam. Food Technology 10: 477-481.

Doyle, M.P. and Marth, E.H. (1975). Thermal inactivation of conidia from Aspergillus flavus and Aspergillus parasiticus. Journal of Milk and Food Technology 38(2): 750-758.

Driessen, F.M. and Bouman, S. (1979). Growth of thermoresistant streptococci in cheese milk pasteurizers - Experiments with a model pasteurizer. Voedingsmiddelentechnologie 12: 34-37.

European Parliament and Council of the European Communities (2004a). Regulation (EC) No 852/2004 of 29 April 2004 on the hygiene of foodstuffs. Official Journal of the European Union L139: 1-54.

European Parliament and Council of the European Communities (2004b). Regulation (EC) No 853/2004 of 29 April 2004 laying down specific hygiene rules for food of animal origin. Official Journal of the European Union L139: 55-205.

European Parliament and Council of the European Communities (2004c). Regulation (EC) 1935/2004 of 27 October 2004 on materials and articles intended to come into contact with food and repealing Directives 80/590/EEC and 89/109/EEC. Official Journal of the European Union L338: 4-17.

European Parliament and Council of the European Communities (2006). Directive 2006/42/EC of 17 May 2006 on machinery, and amending Directive 95/16/EC (recast). Official Journal of the European Union L157: 24-86.

Food and Drug Administration, Public Health Service (1989). Grade 'A' Pasteurized milk ordinance. Publication No. 229. U.S. Department of Health and Human Services, Public Health Service, Food and Drug Administration, Washington, DC, USA.

Food and Drug Administration, Fish and fishery products hazards and control guide (1997). 1st Ed. U.S. Department of Health and Human Services, Public Health Service, Food and Drug Administration, Washington, DC, USA.

Food and Drug Administration (latest update). Food Drugs Cosmetics Law Reports §178.3570. U.S. Department of Health and Human Services, Public Health Service, Food and Drug Administration, Washington, DC, USA.

Langeveld, L.P.M., Montfort-Quasig, R.M.G.E. van, Weerkamp, A.H., Waalewijn, R. and Wever, J.S. (1995). Adherence, growth and release of bacteria in a tube heat exchanger for milk. Netherlands Milk and Dairy Journal 49(4): 207-220.

Lelieveld, H.L.M., Hugelshofer, W., Jepson, P.C. *et al.* (1992). Microbiologically safe continuous pasteurization of liquid foods. EHEDG Document 1. Campden Food and Drink Research Association, Chipping Campden. (Summarized in Trends in Food Science and Technology 3(1992): 303-307).

Lelieveld H.L.M. (1994). HACCP and hygienic design. Food Control 5: 140-144.

Lelieveld H.L.M. (1996). Protection of microorganisms in soil against thermal and chemical inactivation. In: Proceedings of the fourth Asept international conference Food Safety '96. A. Amgar (ed.). Asept, Laval, pp. 337-343.

Lelieveld H.L.M. (2000). Chapter 59 in 'The Microbiological Safety and Quality of Food' Volume II Part IV. Editors: Barbara M. Lund, Tony C. Baird-Parker and Grahame W. Gould. Aspen Publishers, Maryland, pp. 1656-1690.

Lelieveld, H.L.M., Mostert, A.M., White, B. and Holah, J.T. (eds.) (2003). Hygiene in food processing: Principles and practice. Woodhead, Cambridge, UK, 408 pp.

Lelieveld, H.L.M., Mostert, A.M. and Holah, J.T. (eds.) (2005). Handbook of hygiene control in the food industry. Woodhead, Cambridge, UK, 744 pp.

Lelieveld, H.L.M. and Holah, J.T. (eds.) (2011). Hygienic design of food factories. Woodhead, Cambridge, UK, 750 pp.

Luning, P.A., Marcelis W.J. and Jongen, W.M.F. (2002). Food quality management. A techno-managerial approach. Wageningen Academic Publishers, Wageningen, the Netherlands.

Timperley, A.W., Axis, J., Grasshoff, A., Hodge, C.R., Holah, J.T., Kirby, R., Maingonnat, J.F., Trägårdh, C., Venema-keur, B.M. and Cerf, O. (1993a) A method for the assessment of in-line steam sterilizability of food processing equipment. EHEDG Document 5. Campden Food and Drink Research Association, Chipping Campden. (Summarized in Trends in Food Science and Technology 4: 80-82).

Timperley, A.W., Axis, J., Grasshoff, A., A., Hodge, C.R., Holah, J.T., Kirby, R., Maingonnat, J.F., Trägårdh, C., Venema-keur, B.M. and Cerf, O. (1993b). A method for the assessment of bacteria tightness of food processing equipment. EHEDG Document 7. Campden Food and Drink Research Association, Chipping Campden. (Summarized in Trends in Food Science and Technology 4: 190-192).

Van der Wende, E., Characklis, W.G. and Smith, D.B. (1989). Biofilms and bacterial drinking water quality. Water Research 23: 1313-1322.

Van der Wende, E. and Characklis, W.G. (1990). Biofilms in potable water distribution systems. In: G.A. McFeters (ed.)Drinking water microbiology: Progress and recent developments. Brock/Springer Series in Contemporary Bioscience. New York: Springer-Verlag, pp. 249-268.

Venema-Keur, B.M., Axis, J., Grasshoff, A. (1993). A method for the assessment of in-line pasteurization of food processing equipment. EHEDG Document 4. Campden Food and Drink Research Association, Chipping Campden. (Summarized in Trends in Food Science and Technology 4: 52-55).

Vinseon, H.G. (ed.) (1992). Guide to addenda to contracts within the food industry. Institute of Mechanical Engineers, London.

Verrips, C.T. and Van Rhee, R. (1981). Heat inactivation of Staphylococcus epidermidis at various water activities. Applied and Environmental Microbiology 41(5): 1128-1131.

Wirtanen, G. (1995). Biofilm formation and its elimination from food processing equipment. VTT Publication 251, VTT Espoo.

9. Life Cycle Assessment (LCA): what is it and why is it relevant in food product design?

Chris E. Dutilh
Consultant on Environmental Matters, Amsterdam

9.1. Introduction

As part of their sustainable development programmes, companies usually start by analysing the current situation. In doing so, they identify opportunities for improvement, which can be the start for an improvement process about which they report and which can be evaluated in order to find out what to do next. Environmental management systems are being installed in factories, in order to measure and monitor the environmental impact. Improvement targets can be formulated and subsequently implemented. However, usually only a minor part of the total impact of a product is generated in a company's own operation (Figure 9.1).

In making margarine, for instance, raw materials are put together, packaged and subsequently sent to a distribution centre. For this process energy is used, and waste(water) generated. But the raw materials have been produced elsewhere, and the products are transported from the distribution centre to a supermarket and next to a household, where they generate their own environmental impact, like energy use for refrigeration, contamination of tap water by detergents during the washing-up and where packaging waste is left. So when looking at the overall impact of a margarine, it becomes clear that only part of its impact (between 10 and 20% of the overall effect) is generated within the manufacturer's direct control. The remainder occurs elsewhere (between 80 and 90% of the impact).

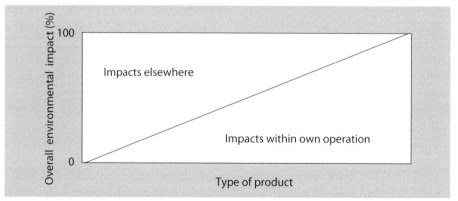

Figure 9.1. The overall environmental impact of a product is partly generated within a manufacturer's own operation. The remainder is generated either up-stream in the production of raw materials or down-stream in using the product.

Most companies focus on the impact they generate themselves, at least in the beginning, which is very logical, because improvements in the lower triangle (Figure 9.1) usually generate money or avoid the waste of money. Hence at the 3M company they say: 'pollution prevention pays'. However, in the upper triangle companies can also make a difference. A company can significantly influence the overall impact of its products by selecting raw materials and suppliers which generate a lower impact and by formulating the products in such a way that they do not generate impacts elsewhere.

Life Cycle Assessment, or LCA, is a method that enables a company to compare the overall environmental impact of products and services 'from the cradle to the grave'. Governments all over the world are encouraging the use of LCA. For example, LCA is one of the core elements in the development of environment policy in the European Union and the European Commission is supporting both businesses and policy makers to use the tool. LCA has become a standard tool for companies throughout the world to investigate the environmental impact of their products. It is important for businesses and authorities to know how LCA is implemented and how to communicate its results to suppliers, customers and public stakeholders.

This chapter aims to clarify the method, tell you how it can be applied, and what can be done with the results. It will help you understand some of the benefits and limitations involved in using LCA. The second part of the chapter ('How to carry out an LCA, step by step') provides a brief description of the method. If you want to start using LCA yourself, you will find suggestions for further reading. If you are mainly interested in the results, we have added some examples to show the kind of information a LCA study can produce and what you can do with it.

9.2. What is LCA?

9.2.1. Strengths and limitations

Life Cycle Assessment (LCA) is designed to identify the impact a product or service has on the environment throughout its entire life cycle. It is based on a systematic examination and evaluation of the environmental impact. This impact is assessed both in terms of the use of raw materials and space and in terms of emissions of potentially hazardous substances (such as CO_2, SO_2, phosphates, heavy metals and pesticides), noise, radioactive substances, etc. Factors which are not directly environment-related, like product safety, cost and social factors, are not considered. An LCA study only gives information about the environmental dimension of sustainability. In some Nordic countries in particular more extensive forms of life cycle assessment are being developed, which also consider the economic and social dimensions of sustainability.

Broadly speaking, a life cycle consists of all the processes connected with the production and use of a product: from the extraction of the raw materials, right up to reuse, recycling and final disposal. It is possible to analyze a specific product, or a service in which one or more products are used. For example, LCAs are used to compare the use of glass bottles with the use of tins, plastic fibre with steel cable, or transport by road with transport by rail.

9.2.2. Customer requirements

Increasingly, manufacturers are using a problem-based approach to identify ways of improving production processes, embracing almost the entire supply chain. An entry check on raw materials is no longer sufficient. Customers want better registration to be able to determine the origin and components of a product. When buying a product, they do not want to risk being held liable for something from an earlier phase of production.

9.2.3. What can you expect of it and what not

LCA is a useful instrument for identifying the environmental impact of products from the cradle to the grave. Because it addresses the entire life cycle it makes it easier to tackle problems rather than duck them, for example by transferring the impact to a later stage, or from one issue to another (less waste, but more energy consumption) or from one phase in the life cycle to another (for example from production to consumption). Apart from LCA there are other tools, such as risk assessment, material flow analysis, environmental impact assessment (EIA) and environmental audits, each with its own specific area of application. EIAs, for example, are used to compare locations for a new building or activity.

LCA is not a 'supertool', and LCA studies do not in themselves solve problems. They do, however, provide information which can be used as the basis for decisions and environment policy measures. In this sense LCA is a decision support tool, to be taken into account in addition to other important considerations, such as technological, economic and social factors.

Perhaps even more important than the LCA study is the concept of 'life cycle thinking'. After you have carried out several LCA studies you generally automatically start thinking in life cycle terms, which means you can avoid ducking issues at an early stage. Sometimes this produces ideas for totally new products and processes, such as a clockwork radio, which was developed after discovering that by far the greatest environmental burden in the life cycle of a radio was caused by the batteries.

9.2.4. From a historical perspective

The first LCAs date from the mid-1970s and are basically detailed energy and waste analyses. In the early 1990s LCA was in the news because of a debate about the relative environmental merits of cotton versus paper nappies and of washable rather than disposable coffee cups. These discussions highlight both the strengths and the limitations of LCA. It is at its best then on the search for improved products or processes. Or, where similar alternative products exist, LCA can be used to compare them. It is less suitable for comparing products which are highly dissimilar, as there is no absolute measure for the environmental burden of a product. Where opinion is divided, LCA provides a convenient starting point for discovering where there is common ground and on which issues there is still disagreement. In that sense, too, LCA operates as a decision support tool.

9.3. Who can benefit by it?

9.3.1. Businesses

The outcome of an LCA study is relevant to many people in your business. Buyers gain an idea of the standards they can require of suppliers, and product developers learn which factors contribute most to the environmental burden. An LCA study makes the environmental burden of your products visible, and this can be useful:

• when you are faced with critical customers, consumers, public authorities or environmental interest groups. With the facts at your fingertips you are able to respond far more adequately. Whether you feel this is important or not depends on your line of business. Because industries such as building, food, personal hygiene, packaging and plastics regularly face criticism, quite a few companies and umbrella organizations have started using LCA;
• LCA enables you to provide environmental data about your products to companies you supply. This is increasingly proving a competitive advantage;
• LCA enables you to employ environmental arguments in your communication strategy. For example, an LCA study can help you decide whether it is worthwhile applying for an eco-label or environmental grant;
• LCA gives you a firm foundation for strategic and design decisions when developing product policy. Incidentally, it is worth knowing that LCA studies often highlight ways of operating more efficiently, thus generating cost reductions;
• of course, if all else fails, you can use LCA to try and prove that your product is more environmentally friendly than that of your competitors, though this is one approach that very rarely works.

In practice your Environment or R&D Department will usually carry out the LCA studies or coordinate contacts with LCA consultants. Often companies do not communicate the results of LCA studies as such, but generally publish them on the basis of specific themes, such as energy consumption or use of resources.

9.3.2. Consumers

Individual consumers and consumer organizations are also interested in the 'world behind the product', the environmental burden a product has caused before it reaches the shops as, for instance, illustrated by 'the StoryOfStuff (www.storyofstuff.org)'. This is exactly what LCA can offer: all forms of environmental burden identified and totted up, presented clearly and transparently. The average consumer is not interested in a fat file full of detailed findings, but wants to take in the information at a glance, for instance on a label. Consumer organizations have different information needs. They include the environment as a quality aspect in their comparative tests. The more they can rely on LCA, the more businesses, public authorities and consumer organizations will share a common language. This will prevent opinions forming on the basis of myth and prejudice.

9.3.3. Public authorities

Public authorities are also increasingly interested in comparing various policy options in a broader context. For example, in policies designed to achieve a steady reduction in the environmental burden of products. Government incentives for green investment and green purchasing are also supported by insights based on LCA. Government departments in the Netherlands have carried out several LCAs to establish the most environment-friendly way of building infrastructure. The national waste management platform has compared various waste disposal scenarios using LCA studies. LCA has also been used to include environmental issues in the public debate on building and packaging. Many of these initiatives are encouraged by the government, particularly because they give a better insight into the scope for environment-related product improvement, but also because they quantify the consequences of alternative policy options.

9.3.4. Students

Students on various courses benefit by becoming familiar with LCA at an early stage. This applies to budding designers, who thus learn to fully appreciate the environmental consequences of (adapting) their designs. But it applies equally to students of nutrition, who quickly learn to understand the consequences of different patterns of production and consumption for the environmental burden in various phases of the food chain.

9.3.5. The international arena

On the European level, the European Platform on life cycle thinking and assessment provides consistent and quality-assured life cycle data, methods and assessments. This European Platform also hosts a selection of tools, reference data and recommended methods for LCA studies (http://lct.jrc.ec.europa.eu).

Internationally, the United Nations Environment Programme (UNEP) is the main player in the field of LCA. The UN wants to encourage the application of LCA all over the world, particularly in developing countries. But these countries feel the LCA approach is still a matter for the rich, industrialized nations, whose calculations take insufficient account of environmental effects outside their borders. This is why authorities in every country and every company in the world should know about the way LCA is implemented and how to communicate about it with suppliers, customers and other stakeholders. Meanwhile a series of ISO standards and technical reports for implementing LCAs have been developed (ISO 14040 series). Moreover scientists, industry and governments are working together for the further development of the LCA method in platforms such as SETAC and Spold (www.spold.org). Together with SETAC, UNEP has taken the initiative to improve international co-operation. More information on this Life Cycle Initiative can be found at their website (www.uneptie. org/sustain/lca/lca.htm).

9.4. How to start an LCA study

If you want to start an LCA study you can:
- consult others who already have experience conducting an LCA study;
- get in touch with a consultancy to carry it out for you;
- search for more information about the method in booklets, on a course or on the Internet;
- simply do it yourself on the back of an envelope.

9.4.1. From LCA studies to integrated life cycle thinking

The first LCA study is often time-consuming, mainly because you have to set up a database with information that is of specific relevance for your product and the related production processes. You are generally highly dependent on information from third parties. A great deal of information is already available for many standard processes. In any case it is important to make sure that the available information is sufficiently representative of the process or product you are analyzing. Once you have built up a database and gained some experience from previous projects you will find it increasingly easy to carry out follow-up studies. You will also be able to review your previous studies, basing them on better information. In fact there will be a transition from isolated LCA studies to integrated life cycle thinking. Of course you will have to continue to maintain and update your database.

9.4.2. Whether to use a consultancy or do it yourself

Because the studies are complex and the basic data is not usually readily available, businesses generally use a consultancy to carry out an LCA study. Even so, there are some (mostly larger) companies that recognize the advantages of carrying out their own LCAs:

- Much of the information is confidential and therefore more easily accessible to the company than to a consultancy.
- It is much easier for a manufacturer to judge the relevancy of a production phase, which saves looking for information on the environmental burden of less relevant phases.
- The knowledge gained can be put to immediate use.
- It is often easier to control the cost of a study if you carry it out yourself.

One disadvantage of carrying out a study yourself is that it might be felt to be less objective, which may be an important consideration.

9.5. How to carry out an LCA, step by step

9.5.1. General approach

As with any study, the first thing to do is define clearly the purpose of the LCA and the audiences the results are intended for. When these elements are clear, you can start setting up a flow diagram. In the flow diagram you set out the phases in the production process of the product you want to examine. You then estimate the impact on the environment of each of these phases. Finally, you may want to manipulate the information you have obtained to get manageable indicators.

Clearly the results of an LCA study should be useful and acceptable to the intended audience. It is important to consult all interested parties on what to examine if you want to carry out an efficient study with useful results. It is often worth involving these people in carrying out the study. You should agree clearly defined procedures, particularly when you want to use the results in a public debate. You can promote participation by inviting the parties to join a focus group where they can discuss findings and dilemmas.

The strength of the LCA method is its systematic approach. In general, four phases are identified:
1. goal and scope definition;
2. inventory analysis;
3. impact assessment;
4. interpretation.

Each of these phases is subdivided into a number of steps which will briefly be described below.

9.5.2. Deciding goal and scope

The first phase of an LCA study is crucial for the rest of the study. Which is why it is very important that you spend enough time on it.

Functional unit

Does the study address an internal need to identify product improvements, or does it involve the comparison of very diverse products and services? In either case the choice of what to compare is crucial. For example, if you want to compare alkyd paint with acrylic paint, it is no good comparing them in units of a tin, a litre, a kilogram or a euros worth of paint. After all, you need much more of one type to obtain a coat of paint than of the other, the number of coats required for a good finish differs, and one type is more durable than the other. By combining all these factors we end up with a unit like 'one square metre of wall surface, adequately protected for a period of ten years'. This 'functional unit' determines the quantity of paint, paintbrushes and paint remover that have to be compared in the assessment. Though many LCA studies do not explicitly involve a comparison but are designed to discover the principal factors that contribute to the environmental burden of a product, it can still be useful to establish the functional unit of a product.

Alternatives

Apart from determining the functional unit you have to determine what alternatives to involve in the assessment. This generally depends on the object of the study. If you are comparing modes of commuter transport, there is no need to include air transport or high-speed rail in the comparison. Similarly, if you are comparing ways of getting to a Mediterranean holiday destination, there will not be much point in considering the advantages of cycling.

Methodological issues

Before you start to gather information, you should establish what are likely to be the most important methodological issues. These generally concern the choice of system boundaries (what to include and what to exclude). You will also have to think about the method of assessment and assigning the inventory results. By discussing this with all parties beforehand you will reduce the risk of disagreement during the study and of being accused of 'working towards a result'. Often you can avoid disagreement by carrying out the calculations in several different ways right from the start. This quickly shows how sensitive the assessment is.

9.5.3. Inventory analysis

In the second, most time-consuming stage of an LCA study you have to establish the life cycle of the product or product alternatives, first qualitatively and then quantitatively. During this stage you will benefit most from the use of a computer, particularly because of the many phases a production process can involve and the sometimes large number of environmental issues in each phase. Such a flow diagram is called a process tree (see Figure 9.2).

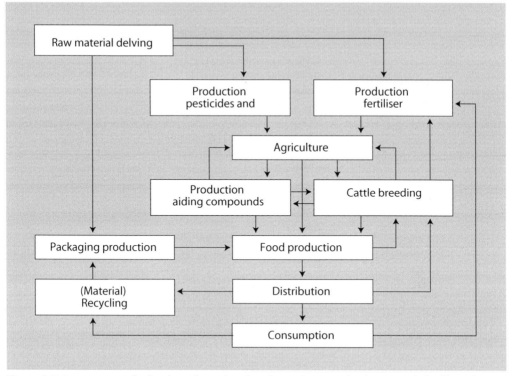

Figure 9.2. Process tree of a food product, showing the various production steps in its life cycle.

Qualitatively

Inventory analysis is based on individual phases in a production process, such as rolling steel in steel manufacture, power generation in a waste-to-energy plant, pressing a CD or recycling paper. Sometimes a whole refinery will be taken as one phase in the production process, while in other cases it might be split into ten separate steps. Together, all the production phases, plus transport, resource consumption and waste disposal constitute the product system. A product system is graphically summarized in a flow diagram composed of all the individual processes.

Quantitatively

Data concerning each process is collected. This is data about categories such as:
- economic input (e.g. power consumption, use of resources, use of recycled materials);
- output of products and any by-products (e.g. electricity, materials, materials that can be recycled);
- input from the eco-system (e.g. crude oil, ores, trees);
- emissions into the eco-system (e.g. CO_2, heavy metals, benzene).

Depending on the nature of the study you can either average this data across a large number of manufacturers or, by contrast, compare data from specific production locations. Experience shows that the amount of data you need to collect can be greatly reduced once you have carried out a few studies, as the environmental burden tends to concentrate on a limited number of categories. Thus, for an electric light bulb, it soon turns out that electricity consumption has the biggest impact, which means you can concentrate on that. However, it is important to check regularly whether the assumptions you based your simplification on are still valid. Once you have described the characteristic environmental mechanisms and their extent for all processes, you have to set appropriate scales.

Collecting data

It is generally not easy to obtain useful data. Manufacturers often do not have the specific data you want. It is worth examining with them whether it is necessary to carry out special measurements to obtain the information or whether it can be found in another way. Sometimes manufacturers do not want to release specific data because it could be used to discover trade secrets. In such cases it can be worth applying for data at industry level or bringing in an intermediary who can be trusted to look at the data and process it on the basis of a non-disclosure agreement. Manuals, technical encyclopaedias and earlier LCA studies on similar products are important sources of data, and more and more information can be found on the Internet.

Estimates

It can prove very difficult to find specific data on a number of processes, either because the information does not exist, because it is confidential, or because you are unable to find the right source. However, you cannot avoid taking these processes into consideration, especially when it is totally unknown how important the environmental impact of these processes is for your study. In that case, you will have to estimate. You can base such estimates on the environmental impact of similar processes or on a so-called input-output analysis, in which the environmental mechanisms are estimated based on data about an industrial sector. Based on your estimate, you can decide whether it is still important to look for more specific information.

Choices

When you have mapped out all the process steps in a flow diagram and have tried to find or estimate the relevant data, it may turn out that for a number of processes you still do not have enough information. This choice can influence the final result and is to some extent subjective, depending on how much time and money you have for the study. When you have made such a choice several times, it becomes easier to appreciate the consequences and to make an informed decision about an appropriate cut-off point. It is important to discuss these choices with everyone concerned in good time.

Multiple products

There is one complication that has to be mentioned. Many production processes result in more than one product. Milk production, for example, is inextricably linked with the birth of a calf, and thus with meat production. Moreover, cows produce not only milk and meat, but also leather and raw materials for the pharmaceutical industry. Somehow the 'environmental interventions' connected with the production of all these products (for example, the production and use of animal feed and electricity) have to be distributed over the various products. A similar approach is needed for processes at the end of the life cycle, where products are brought together for disposal or separated for recycling. This distribution of environmental mechanisms is known as 'allocation'. There are various bases which can be used for allocation: mass, energy content or economic value of the various flows. The latter is generally the preferred method in the Netherlands.

9.5.4. Impact assessment

The outcome of an inventory analysis is an 'environmental intervention table', which consists of a large number of resources and emissions and possible types of land use, expressed in quantities per functional unit. Often this is sufficient for a comparison of alternative products or to suggest approaches to redesign. Sometimes, particularly when there are differences in dissimilar interventions, it can be worth carrying out further assessment. Interventions are, as it were, translated into a list with contributions of those interventions to impact categories such as resource depletion, global warming and acidification.

Environmental profile and impact score

The first thing is to establish the relevant impact categories for the product you are examining. Categories which are often considered include: depletion of energy resources and mineral ores, global warming, depletion of the ozone layer, toxicity, land use, acidification, over-fertilization, smog and odour nuisance. An inventory table may contain ten or more heavy metals (lead, mercury, chromium, cadmium) or other substances which are to some extent toxic. It may also contain a number of acidifying emissions or emissions known to contribute to global warming. For each intervention, its contribution to the relevant impact categories is looked up in special tables ('classification'). Similarly, for each environmental intervention, you look up how toxic, acidifying, etc., the relevant emissions and raw materials are ('characterization'). Finally, the contributions of all interventions in each impact category are added up to produce impact scores expressed in terms of their own unit, like 'kg CO_2 equivalents' for global warming. Taken together, the impact scores produce an 'environmental profile', indicating the size of the aggregate environmental impact of the product, for example with respect to toxicity, acidification, and resource depletion.

Normalization and weighting

The next step may be to relate impact scores to the total annual size of an environmental category, so that they can be expressed as 'the contribution of a given intervention to the total environmental burden in a certain environmental category in a year' (normalization). Standard computer programmes which contain much of the necessary information are generally used for classification, characterization and other calculations such as normalization. Many people will want to simplify the environmental profile even further by weighting the various impact scores and adding them up to obtain one or more eco-indicators. In order to do that it is first necessary to establish the relative gravity of the various environmental categories, and this is a highly subjective matter. That is why this operation usually generates a great deal of debate.

9.5.5. Interpretation

The final phase of an LCA study is to test the results for reliability and validity and to examine the conclusions to see whether they will stand up to scrutiny. In this phase you assess the outcome in the light of the data used. If an environmental category emerges particularly strongly from the evaluation, it is important to discover what production processes were relevant and how reliable your data was.

For more sensitive analyses that you will also want to use externally, it is advisable at this stage to put the results before a group of independent experts for 'critical review', in which they can assess whether your choices were sound and the information you used was sufficiently representative. In fact it is more sensible to approach these 'critical reviewers' at an earlier stage of the study, to consult them on how best to proceed. This can save a great deal of time and irritation.

9.6. Illustrative case-studies

9.6.1. Case study: Single-use or multiple-use packaging

LCA studies have been carried out in the Netherlands since the early 1990s in connection with the debate on packaging. These studies were mainly designed to gain quantitative information about single-use and multiple-use packaging. Not only environmental aspects were taken into account but also the economic implications of a possible transition. Because this was a sensitive issue, interested parties were closely involved in the process. Independent experts critically reviewed most of the studies. In 1994 a number of general conclusions were drawn from these studies, which had wide public support. For example, refill packs for detergent proved a better alternative than refillable packaging. At the same time many of the deposit-and-return options proved so expensive that the environmental benefit, if any, would not have been cost-effective. In 1995 the environment minister accepted the conclusions, which underlines the potential of a carefully conducted assessment. However, this approach does not offer any guarantees for continued acceptance of the results, as became clear when the subject of deposit-and-return was put back on the political agenda five years later.

9.6.2. Case study: LCA provides basis for Dutch eco-label criteria

When developing the diagrams for the Dutch eco-label scheme ('Milieukeur'), LCAs are used to identify focal areas. In general, a consultant will carry out an LCA for a representative product from a given category. The study will indicate which factors contribute most to the environmental burden for that category of products. In addition, it will examine whether there is much difference in environmental burden between the products currently on the market. A board of experts bases the criteria a product must meet in order to obtain the eco-label on this. The criteria are set in such a way that 20 to 50% of the products will in principle qualify for the eco-label. Here the LCA serves as an underlying source of information for the board of experts.

9.7. Further information

The website of the European Platform on Life Cycle Thinking and Assessment, which is an initiative of the European Commission, provides extensive references to further information on data and tools, see http://lct.jrc.ec.europa.eu.

9.8. Some LCA jargon

Allocation	Distribution of process data across relevant outputs of that process
Classification	Qualitative assignment of environmental interventions to the selected impact categories
Characterization	Quantification of the contribution of an environmental intervention to various impact categories
Critical review	Evaluation by an independent expert of, among other things, the transparency and consistency of the assessment
Environmental index	An environmental profile for materials or process phases, based on a life cycle approach and usually reduced to a single figure by means of a specific weighting procedure (e.g. Eco-indicator)
Environmental intervention	Human operation that has consequences for the environment, such as resource depletion or emissions into the air or water
Environmental profile	Table showing impact scores for selected impact categories obtained in a characterization step

Chris E. Dutilh

Flow diagram	Graphical representation of a product system (also process tree)
Functional unit	Basis for comparison of alternative products satisfying the same need
Impact assessment	Third phase of LCA, in which size and importance of selected impact categories are determined
Impact category	Representative category to which a number of environmental interventions can be attributed (also environmental theme)
Impact score	Numeric outcome of characterization for a given impact category
Interpretation	Fourth phase of LCA, in which conclusions are drawn in relation to previously determined goal and scope of study
Inventory analysis	Second phase of LCA, in which relevant input and output data are collected for all elements of the product system
Life cycle	Series of related phases in the life of a product system, from resource extraction to final disposal
Normalization	Comparison of impact scores with reference values
Product system	Group of interrelated production, transport, consumption and waste processes, which together constitute one or more defined functions
System boundary	Boundary between the product system being examined and the environment or between the product system and other product systems
Weighing	Element of impact assessment which makes it possible to aggregate impact scores
Weighing factor	Subjective value indicating the seriousness of impact category

Acknowledgements

The initial version of this chapter was produced in close cooperation with Mark Goedkoop (PRé), Jeroen Guinée (CML) and Pieter Lanser (ENCI). Arjan de Koning (CML) has been very helpful in formulating the current version. The author can be reached at chris@dutilh.com.

10. Managing knowledge for new product development

Peter Folstar
Product Design and Quality Management, Wageningen University

10.1. The role of knowledge in the food industry

10.1.1. Knowledge and value creation

Knowledge has become a key factor in competitiveness in virtually every industrial area including the food industry. In today's environment companies cannot count on a sustainable competitive advantage, but must continuously develop in new directions (Drejer and Riis, 1999). Creating, building and maintaining knowledge are seen as key assets to boost growth and earnings. Jonash and Sommerlatte (1999) concluded that there is a positive correlation between the long-term development of the market value of a company and the commitment to building knowledge: more innovative companies enjoy a share-price premium higher than their less innovative counterparts. Also the authors concluded that the importance of innovation has significantly increased over the past decade.

Knowledge has been the key contributor to wealth creation in the post-Second World War period. In the sector of health, food and agriculture the increase in particular of the understanding in biological and physical sciences, including the ability to process, use and apply the huge amount of data and information, has been seen as a major development behind this wealth creation. It has encouraged an interdisciplinary scientific approach as opposed to the traditional emphasis on disciplines which very often led to fragmentation. The exchange of information on a real time-basis and the creation of global networks have been important elements in making a more modern interdisciplinary approach feasible.

Neumann and Thomas (2002) emphasized that despite the technology revolution, methods for capturing, assembling and sharing knowledge in the life sciences are human-based and in the mind of scientists. It has been recognised that there is an increasing need for computer-based systems to support the logical interpretation and synthesis over a wide variety of domains. Knowledge assembly systems conducted on information from multiple distinct sources have been recommended. The need to improve on information management has also been recognised in food development (Benner *et al.*, 2003). Much of the scientific effort that has gone into creating new approaches for information and data management will determine the rate at which knowledge-based value creation will take place, also in the food industry.

On the basis of surveys that have been conducted in the life sciences-based industry, the following roles of knowledge are clearly recognised (Folstar, 2001). Knowledge is a *source of excellence*, a cornerstone for a company culture, directed at being outstanding in benchmarking with competition and generating unique products and added value. Knowledge is clearly seen as a *creative motor*, as a driving force behind innovation. It is also the basis of a company's reputation: dedication to research puts the company in the top league of outstanding businesses. Knowledge also enables a company to secure *continuity* in the portfolio of products.

On a more practical level, it enables the company to deal effectively with *claims*: this is highly relevant in the food industry, where safety-related problems require immediate access to up-to-date knowledge. As mentioned before, knowledge also improves the *market value* of a business. From a human resource point of view a knowledge-based company culture attracts high-level professionals, who can be seen as key to *management development* and will help to secure senior positions for the future. On the basis of their knowledge, life sciences-based companies in particular develop attractive portfolios of *patents*. These patents are not just a means of solidifying the future from a defensive point of view, but are also a source of *licensing income* in strategic partnerships. Finally, knowledge is seen as an asset for building internal and external *new networks*, one of the most important reasons behind success in innovation.

In spite of the overwhelming amount of literature on innovation and recommendations for improvement of the innovation process, Stewart-Knox and Mitchell (2003) reported that the vast majority of new food products (72-88%) continues to fail. If a 'new' food product is defined as 'one that is new to the consumer' only 7-25% of food products launched can be considered truly novel. The authors explain that only very limited literature exists on issues of innovation in - specifically - food product development. From this literature, food product success appears contingent upon a high-quality product, senior management support, sound knowledge of the consumer and cross-functional teamwork. In a model specifically developed for low-fat food products the authors emphasize the importance of knowledge of consumer requirements and solid technical capabilities anchored within the company and made operational in interdisciplinary teams.

Mark-Herbert (2004) focuses on the need to identify critical technologies and recommends new strategies for collaborative R&D activities. Referring to the innovation potential in the food industry in Germany, Menrad (2004) sees implementation deficits with regard to market research, especially in smaller companies. The author emphasizes the need to actively integrate new technologies, and also sees a major limitation here to smaller companies and recommends a reshaping of funding mechanisms behind public-private collaboration to facilitate the transfer of technologies.

Although the interest in knowledge for food innovations is growing, it seems that there is still a wide gap between such expertise in the food area and other industrial areas. This chapter will review some of the trends, implications and directions for optimizing the use of knowledge for food innovations.

10.1.2. Trends in the food industry

The worldwide market for processed food grows 2-3% annually (Roels, 2000). Today's food industry focuses on growth and improvement of profitability through innovation and brand development. Marketing and advertising have become dominant activities and the satisfaction of consumer demands forms the basis for successful products (Karel, 2000). Lifestyle and fashion are determining consumers' preferences: today there are opportunities in the area of healthier and more convenient foods. This has already led to unique groups of functional foods, organic foods and culinary foods. An entirely new service industry of catering and restaurants and new product concepts, which are easy to prepare, have also emerged.

Over a number of decades the food industry has developed an ability to organise itself as a part of larger networks in the form of food supply chains, interconnected systems with a large variety of complex relationships, driven by the reality that demand is no longer confined to local or regional supply. As a part of this much emphasis has been placed on quality and safety concerns as well as on environmental issues (Trienekens, 2004).

In our own research we developed a model (Figure 10.1) that addresses an important element in food innovation: the ability to exchange information within a food supply chain between raw material suppliers, ingredient suppliers, food manufacturers, equipment suppliers, packaging material suppliers and retailers focusing on the creation of value. Traditionally, each of these stakeholders had a rather isolated position, but today the need for cooperation throughout the production chain is perceived as leading to more competitive advantage. There will be an increasing

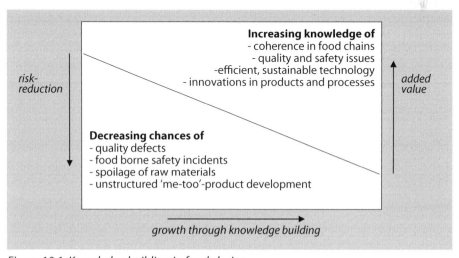

Figure 10.1. Knowledge building in food chains.

trend towards sustainable value creation at the food supply chain level in addition to innovation at the individual company level.

10.1.3. Trends in food science and technology

It is believed that the scientific knowledge and understanding of consumer preferences and the potential to satisfy those through food technology will increase over the next decade. According to De Rooy (2000) this will require an interdisciplinary approach focusing on nutrition, food functionality, sensory attributes and consumer preferences including all the sciences behind these areas. These sciences are in the first place the more 'established' sciences of physiology, biochemistry and biotechnology, molecular biology and process technology. However, other sciences will also enter the food laboratory such as psychology, neurosciences at the one end and information technology and mathematics at the other. Due to the trend towards an interdisciplinary approach, innovative forms of collaboration inside and outside the food industry will be found (Karel, 2000). De Rooy (2000) recommended a systems approach that has the ability to assess all relevant consumer items and product attributes within a given concept.

As mentioned in 10.1.1 science is the backbone of knowledge-driven innovation. The industry requires a continuous injection of knowledge to fuel sustainable growth and new product development. For individual companies this requires a match between the business strategy and a selected scientific profile. This profile is seen as one of the keys to commercial success: in order to derive the full benefit, other factors such as an open, dynamic culture, access to sufficient critical mass, a constant drive towards excellence, interdisciplinary cooperation and full integration of science in business driven innovation need to be fully recognised as a part of strategic leadership (Ganguly, 1999).

In a 2003 forecast collected from leading groups within the Institute of Food Technology (Mermelstein, 2003) a wide range of advances from long-term to short-term applications was foreseen. There is no doubt that the study of functional genomics, the knowledge of complete genomes of plants, animals and micro-organisms and the availability of high-throughput analysis and DNA chip technology will affect the direction of food research in many ways. A better understanding of nutrient-gene interactions will offer a significant opportunity for new food product development focused on the individual in order to lower disease risk. Functional genomics will establish methods to be used to develop food tailored to those who are suffering chronic diseases and will create new opportunities for preventing food-related diseases such as obesity and diabetes (Müller and Kersten, 2003).

In food safety new more rapid and more specific methodologies from molecular biology will become available. Similarly advances in preservation, for instance through naturally-occurring preservatives and the possibility to produce these using modern biotechnology, will increase the opportunities for new product development. Continued improvements in processing and packaging will also increase the feasibility of diminishing food safety risks while, at the same time, creating opportunities for fresh, easy-to-prepare products. Furthermore, advances in areas other than molecular biosciences are expected to make long-term contributions, such as material technology for lab-on-a-chip concepts and information technology for computer modelling of safety risks as well as new complex product development schemes (Baeumner, 2004).

Understanding food preferences has challenged consumer scientists and sensory scientists to improve and develop new methods, in particular for understanding the subjective element in consumer preferences and correlating organoleptic properties with a chemical composition and physical structure: the impact of sciences that are relatively new to the food area such as psychology and neurosciences as well as medical technology applications will most likely lead to breakthroughs (De Rooij, 2000).

There are countless other examples. However, the objective of this chapter is not to be conclusive, but rather to recognize trends that might lead to breakthroughs. As far as the contribution of research to business is concerned, it is expected that such breakthroughs are more likely to occur if:
1. scientific excellence in individual disciplines is encouraged;
2. interdisciplinary projects focused on jointly agreed targets are effectively established, and
3. an open and dynamic culture exists, which is able to absorb new science, and in particular science that is new to the area of food science and technology

10.2. Basic principles of knowledge management

10.2.1. Definitions

Knowledge management is a broad term, which includes all the human and organisational aspects of integrating knowledge as a basis for the creation of added value in an organisation. Knowledge is not just about science and technology. It encompasses all information, experiences and skills, either stored in people's heads (*tacit knowledge*) or stored in media external to people (*explicit knowledge*). Although explicit knowledge is most visible in specifications, recipes, procedures, reports and patents, tacit knowledge - once made explicit - probably offers more possibilities for competitive advantage. Tacit knowledge finds its basis in expertise, skills and creativity, and is therefore also intangible and volatile; but if used well, it can be indispensable to an organisation (Cook and Brown, 2002).

Knowledge management is an activity aiming at improving the innovative position of a business. As will be explained in 10.2.3 *innovation* is the generic business activity aimed at creating competitive advantage by involving all knowledge and expertise within the entire company across traditional barriers (Tidd *et al.*, 2001).

The role of *research* is more explicitly aimed at the exploration of new knowledge in an orderly fashion. Therefore, research is at the basis of processes exploring the potential role and contribution of science to a business.

In order to be relevant, however, a *development* step is needed: this is the process by which knowledge emanating from scientific research is able to deliver goods or services, by using an appropriate vehicle of technology.

Therefore, *technology* is defined as the means by which scientific discoveries are converted into products.

Technology platforms represent larger entities of such expertise and knowledge; as explained in 10.3.3 these platforms represent structures of a certain minimal critical mass where all knowledge and expertise in a certain area ('we know how to...') is brought together.

The terms defined in this paragraph are sometimes used in a less rigorous way and are therefore not entirely representative of their original role. A clear definition is therefore important (Ganguly, 1999).

10.2.2. Conditions

According to Tidd *et al.* (2001) an innovative culture requires management effort in five key areas: strategy, process, resources, organisation and learning. This approach has significant implications for the role and contribution of knowledge: not as a single item in a linear chain approach, but as an integrated function in networks and teams with a dedicated goal. Knowledge management is therefore adept at recognising the potential of any piece of information, any innovations and any new skill that arises.

This may include elements such as customers' suggestions for product improvement, such as suppliers' contributions to process improvement, and scientists' views on entirely new approaches to traditional products or processes.

Each successful knowledge management system is built on five pillars (EIRMA, 1999):
- the organisational structure of knowledge management and the way it has been embedded as part of the overall structure of the organisation;
- the behavioural aspects of knowledge management, the attention to the power of the organisational culture and the role of human resource management;
- effective tools that help to implement knowledge management, to make regular assessments of strengths and weaknesses and to address opportunities;
- the need to measure the effects of knowledge management through realistic goals;
- an action-oriented culture as a driving force behind processes and projects.

The effectiveness of knowledge management depends to a great extent on how the elements mentioned in this paragraph have been embedded in the overall strategy and culture, and it has a cyclical character (Figure 10.2). Top-down vision and support are indispensable and create the basis for knowledge goals, an analysis of opportunities and benefits, a review of financial and personal resources that should be made available, and finally, well-managed projects with clear objectives and relevance for the strategy of the company.

In order to ensure that the contribution of knowledge is fully used, it is important that senior management creates a culture of inventiveness, communication and sharing as keys to sustainable long-term competitive advantage.

10.2.3. Innovation and NPD

Innovation is a generic business activity aimed at sustainable growth and competitive advantage. Innovation includes everything that goes into the creation of new products, services and processes. It involves the entire company across traditional barriers. According to Tidd *et al.* (2001) innovation as a core business process involves:
- scanning the environment for threats and opportunities for change and collecting signals;
- prioritizing signals on the basis of a strategic view;
- obtaining the resources to enable the response;
- implementing the project from start to finish.

Once implemented as a process in a company's culture, the innovation process, by means of its cyclical nature, will become a source of learning experiences as well.

One such learning experience was reviewed via a worldwide survey by Jonash and Sommerlatte (1999) on the basis of cross-selection of the industry worldwide. They concluded the following:

Figure 10.2. Strategic implementation of knowledge management.

- innovation is seen as a critical business factor despite the fact that few companies claim to achieve a highly effective innovation performance;
- major obstacles were resource related, e.g. skilled leaders and professionals as well as the ability to acquire necessary information through intelligence;
- clear top-management commitment to innovation is a basic requirement;
- innovation requires an entrepreneurial and cooperative spirit: managing such a process requires special skills that need constant attention;
- those companies who measure their performance in innovation are more likely to derive business value over time from their efforts.

A wide range of literature exists on topics related to new product development in a number of industrial areas and the factors that determine its success (Montoya-Weiss and Calantone, 1994; Cooper and Kleinschmidt, 2000). In our group Den Ridder and Vernooij (2002) made a critical review of the most commonly cited success factors for new product development:

- top-management support and commitment to the project, including their day-to-day involvement and their determination to allocate resources;
- technological synergy between the requirements of the project and the technological skills of the company;
- market potential in size and growth, as well as an indication of consumer needs;
- proficiency in translating a product concept into a commercial success;
- organisational skills to manage planning, resources and monitoring of a project;
- customers' perception of product advantage;
- proficiency of pre-development activities such as a detailed market study and preliminary business analysis;
- market synergy between requirements of the project and a company's skill in market analysis and product positioning;
- time-to-market, referring to skills to manage launch time and development cycle;
- proficiency of marketing activities to translate consumer needs into a concept and successfully position the product in the selected market.

The importance of each of these factors changes over time, and is dependent on the nature of the product and the position of the company in its business area.

10.3. Knowledge in industrial environment

10.3.1. Sources of knowledge

Knowledge for innovation is available in different forms (10.2.1) and from different sources. Over the years the food industry has been able to utilize a large network of expertise and science from sources within individual companies, from larger industrial conglomerates and from outside suppliers. Figure 10.3 shows a schematic presentation.

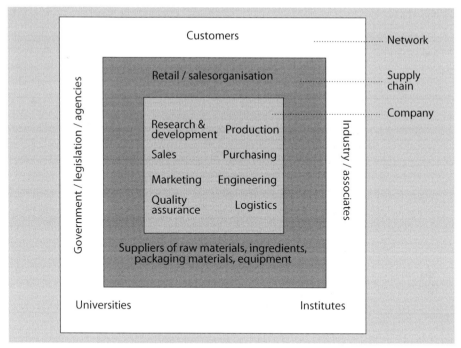

Figure 10.3. Internal and external sources of knowledge in the food industry.

Sources of knowledge include both the existing situation within a company as well as opportunities outside the company, the direct network through various stakeholders in the supply chain as well as a larger, but less pre-determined network of governmental bodies, research institutes and universities and others. Making this knowledge available and operational is one of the great challenges for a knowledge-driven company in the food industry.

10.3.2. Flow of knowledge

In order to understand the strengths and weaknesses of the knowledge level of an organisation, it has been found useful to identify key steps in the organisational backbone of knowledge management. In the light of the goals to be achieved the following steps should be considered (Probst *et al.*, 2002).
- Knowledge identification: this is an attempt to establish a knowledge or competence map ('the yellow pages') that includes all internal and external resources. This exercise, if done in an atmosphere of total transparency, creates an excellent benchmark to competitors and reveals strengths and weaknesses in all competencies on all levels.
- Knowledge acquisition: on the basis of identified weaknesses and/or opportunities, priorities will be established to attract knowledge from external sources. These

may include external experts, other companies, suppliers and other stakeholders, knowledge products including patents.

- Knowledge development: through internal goal setting new opportunities are created and internal resources are made available, for instance in cross-functional teams leading to new products. This may also include a competences development programme on both an individual or group level.
- Knowledge distribution and sharing: once cooperation and team formation is seen as a core value, barriers to sharing should be removed and best practices should be used as examples for changing the culture. These priorities should be encouraged by adequate organisational measures as long as these do not create that common enemy to sharing and cooperation, better known as bureaucracy.
- Knowledge preservation: core knowledge in the form of collective knowledge, key employees, patentable knowledge and data and IT systems should be actively preserved and kept up-to-date.
- Knowledge use: through structuring, but also through best practices and successful new innovations, use of knowledge should be encouraged at all levels within the organisation.

Frequently, progress needs to be measured in the light of the goals established for the knowledge teams. Although presented as a linear model, each of the steps in the knowledge flow makes its contribution as part of a 'living' network.

10.3.3. Implementation

Knowledge is hard to manage: as discussed in 10.2.1 it is present in different forms and is either tacit or explicit. Management of knowledge is certainly supported by (Newell *et al.*, 2002):
- creating the appropriate conditions in terms of culture, focus, communication, finance and infrastructure (10.4.1);
- providing the necessary means in terms of tools for information, reporting, goal setting and processing and applying results;
- creating an action-oriented culture where people are encouraged to seek, share and apply knowledge;
- defining an outside benchmark as a target and bringing in outside peers to review the quality of knowledge and knowledge management.

It has been recommended to capture knowledge in platforms or technologies: these are more-or-less official structures where all knowledge and expertise in a certain area ('we know how to...') is brought together. This enables an organisation to make a continuous assessment of their knowledge position and to monitor and decide which resources should be invested in certain technologies or platforms. Each technology platform is subject to a certain life cycle, including the following phases (Figure 10.4): emerging, pacing, key, mature, aging.

Increasingly, technology platforms are created in partnerships with parties outside the organisation or company. In the food industry partnerships exist on the basis of networks in the supply chain between ingredient suppliers, food manufacturers and retail organisations. Equipment manufacturers or suppliers of packaging materials are often involved in such networks as well.

More recently networks were developed between companies and academic groups in order to create more fundamental research programmes: due to high costs and long development times, these programmes are often co-funded by government because of their pre-competitive character. One example is the network developed by the Dutch Food Industry and leading universities and institutes, the Wageningen Centre for Food Sciences (WCFS) (Case 1).

A technological life cycle analysis is the basis for portfolio management (10.5.2), road mapping technology and product strategies (10.5.3).

10.4. Knowledge management and human resources

10.4.1. Human resource management

In knowledge management the most important factor is the *individual*. In that respect systems and databases are enablers but are not main factors. In general, people will spend time learning, asking questions and listening to those who may contribute knowledge if they feel that they need to learn. A positive culture in the organisation towards capturing and sharing knowledge will accelerate such a process.

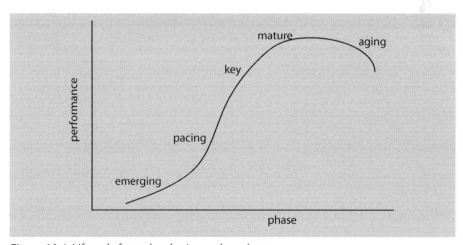

Figure 10.4. Lifecycle for technologies and products.

Case 1: Top Institute Food and Nutrition (TI Food and Nutrition).

TI Food and Nutrition - formerly known as Wageningen Centre for Food Science (WCFS) - is an alliance of research and food industry partners that carries out strategic fundamental research. Organized as a public-private partnership it aims to provide a basis for the development of new food products and thus to contribute to the long-term commercial advantage of the Dutch food industry.

Industrial partners include Avebe, Cosun, CSM, DSM, Unilever and the Netherlands Dairy Industry Association.
Academic partners are Wageningen University and Research Centre, TNO and NIZO Food Research.

Research is carried out in the areas of nutrition and health, structure and functionality and microbiology and safety: project-based research is undertaken at various locations in the participating research organisations. Source: Annual Report WCFS 2003

In this respect Human Resource Management (HRM) has to play a key role: its effectiveness is defined not by its short-term alignment with other management functions, but by its long-term effect on the development of skills, culture and capabilities of human resources. The term 'Human Capital' represents the value that employees bring to their organisation. Measures of this kind of capital are more difficult and context-specific, but progress in accounting for 'intangible assets' is beginning to make their contribution more visible. The effectiveness of HRM for knowledge and innovation has been discussed extensively (Katz, 1997). In this chapter a few relevant issues are highlighted.

With the contribution of HRM there can be a focus on empowering people and stimulating an effective working climate; this could include:
- rewards for building new relationships, not building empires;
- rewards for teamwork instead of individuality;
- encouraging networks of excellence instead of local focus;
- encouraging trust through communication and rewarding requests for help;
- creating space for retrospective learning from successes and mistakes without fear of being penalized for errors;
- rewarding the sharing of knowledge by individuals;
- creating a sense of urgency.

Creating, leveraging and using knowledge are interconnected by a learning curve. The acceleration of learning processes is a cornerstone for effective knowledge management; it will lead to the removal of barriers and finally create a long-term culture of sharing, communication and delivering.

In the management's ideal scenario, knowledge workers internalise the norms, values and goals of the organisation. However, this rarely translates into practice:

engineering a culture from the top down does not work. Norms arise out of shared experience and can be selectively encouraged (Probst *et al.*, 2002).

On a practical level it should be recognized that individuals in knowledge organizations are interested in both the tangible and intangible rewards of their work. As far as tangible rewards are concerned, knowledge workers should have access to the whole system of financial rewards. But perhaps of greater importance are the intangible rewards like status, reputation and recognition by peers and colleagues. Tampoe (1993) summarized four important motivators:
- the opportunity for individuals to realize their potential through personal growth;
- an environment that encourages and enables knowledge workers to achieve their targets and realize their assignment relevance and value to the organization;
- a sense of achievement in accomplishing a task that is of high relevance and value to the organization;
- a financial reward, symbolizing the value of a contribution made by an knowledge worker.

It has to be emphasized that there are both positive as well as negative sides to rewards. Among the negatives are dissatisfaction with others and overemphasis on rewarded behaviour rather than effectiveness. Additionally rewards for individuals may undermine interpersonal trust and make teamwork more difficult. Team-based rewards may offer a solution (Tidd *et al.*, 2001).

10.4.2. Working in teams

Knowledge creation or knowledge generation is typically an activity that is accomplished by a *team* of people rather than by individuals working alone (Tidd *et al.*, 2001). More than anything the development of new products, services or processes typically occurs within teams. In successful teams a number of key resources are present: knowledge of markets and customers, knowledge of available technologies, knowledge of materials, knowledge of distribution processes, all of them usually dispersed over the entire organisation and brought together in a project team or other collaborative arrangement.

Knowledge creation needs to be seen as an interactive team working process; one which involves a diverse range of actors with different backgrounds, cutting across organisational boundaries and combining skills, artefacts, knowledge and experiences in new ways. Thereby trust is perhaps to be regarded as the most critical issue for effective team working and knowledge sharing (Newell *et al.*, 2002). *Creativity* develops from the interaction of different knowledge sets. Occasional contacts are not enough, but must occur in a structured way over prolonged periods.

Unfortunately teams are too often presented as the organisational panacea for all the problems of organisational life: there is a prevailing ideology that espouses the benefits of teams in relation to virtually any organisation's problem including knowledge creation. It has, however, been observed that people fall too readily into patterns of competitiveness, conflict and hostility. Hence, it is often wrong to rush into consensus-seeking solutions and ignore conflict.

Trust is not easy to develop, especially when team members come from different backgrounds with different knowledge and understanding. In such situations, prolonged interaction and common experiences are necessary to develop the shared understanding necessary for trust to develop and knowledge sharing to be possible. Trust is about dealing with risk and uncertainty and also about accepting vulnerability. Where tasks are interdependent and there are material and immaterial matters that one values, vulnerability and the need for trust are higher. In teams conditional trust will usually be converted into unconditional trust, based on experience on both positive and negative occasions. Unconditional trust leads to superior teams and excellence in knowledge creation. Time pressure could have positive effects on *developing trust and coherence in teams*: commonly shared deadlines and team members who have never worked together before can be very useful. In this light it is useful to distinguish between companion trust (emotional, long-term, stable), competence trust (reputation, short-term, more fragile) and commitment trust (formal, contract, micro-term, resilient). It is very important to make sufficient time for developing trust.

High-trust relationships seem to be essential to developing collaboration. Individuals will not necessarily trust each other because they are told to work together in a team. *Trust* is a fertile ground for many knowledge-related processes, for example making knowledge transparent, sharing and distributing knowledge, utilising knowledge. An employee who fears for his job is not likely to share his expertise: he is more likely to try and make himself indispensable. In the absence of trust all talk of an open culture is hollow.

10.5. Operational aspects of knowledge management

10.5.1. Selection and building of projects

Ideally each business has a mix of innovation projects ranging from very short term to more long term as far as the period of accomplishment and contribution to the business goals is concerned. Ganguly (1999) recommends visualizing and managing the selection and building process by using the innovation funnel. This funnel consists of a number of subsequent steps leading from new ideas to implementation and product launch and is represented in its simplest form in Figure 10.5.

Each phase in the funnel is completed on the basis of 'go/no go' decisions at a gate, to ascertain the chances of success. Selection criteria used at gates include a critical review of the question as to whether ideas meet consumers' needs and fit into the business strategy (gate I), the review of the feasibility of the project with regard to

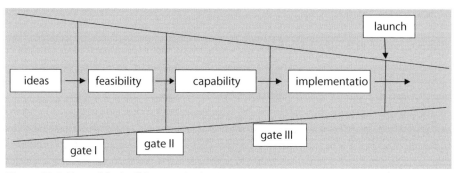

Figure 10.5. Funnel for building and selecting projects.

expected performance, financial viability, marketing criteria and human resources requirements (gate II), and a final review as to whether all elements of the product-marketing mix, capital requirements, key functional areas and commercial viability have been implemented to justify a launch (gate III). According to the author the funnel technique offers applications for both larger innovation projects in general as well as for other research and development projects.

Many companies have adopted a Stage-Gate approach (Cooper *et al.*, 2001) to manage projects through a number of sequential phases: these include idea, concept, project and business. As a project progresses through the stages, the level of information increases and becomes more detailed. At regular Gate Review Meetings go/no-go decisions are made, including decisions on (re)allocations of resources. These decisions are closely linked to a review of portfolios, both on a product development project level and on the level of technology portfolios (10.5.2).

Ganguly (1999) also stresses that a vital part of project management is the capability to assess and manage risks associated with the idea. Although risk management should be an integral part of the innovation management process as a whole, it is of special importance to the feasibility phase. Four generic steps should be included:
1. risk identification leading to an overall review of the possible risks involved;
2. risk assessment leading to a priority listing of individual risks;
3. risk reduction leading to a plan to minimise risks in a worst case scenario as well;
4. risk control leading to management activity that involves regular (re)assessment and identification of new risks.

Managing projects is aiming at 'doing things right' whereas managing a portfolio, which will be discussed in 10.5.2, is aiming at 'doing the right things'.

Defining *goals* is an essential element in knowledge and project management. In knowledge organisations and in the knowledge-based industry goal setting for knowledge is underdeveloped when compared to other business areas such as finance or sales and marketing. Knowledge goals are still in their infancy and suffer from the absence of clear, measurable and concrete instruments. Probst *et al.* (2002) explained that knowledge goals will point towards areas and sources of knowledge that need further strengthening and new areas that need to be created. It will cover both internal and external sources. Identifying gaps can be an effective trigger for a learning process and it will be the basis for a dynamic approach of knowledge and innovation, as discussed in 10.3.2. Creating access to knowledge is increasingly a part of life: success in doing so can lead to significant competitive advantage.

It has also been shown that the development of clear goals against which progress can be measured are very important for relevance to business. A mechanism built on benchmarking with competitors, mainly with groups outside the company, has been found to be particularly effective. Therefore, it is important to reach agreement on concrete, output-relevant goals. A number of goals are given here as examples:
- realisation of a new (group of) products, preferably contributing to a team effort measured as a percentage contribution of the company and/or market share;
- improving the knowledge position in certain (new) areas of science and technology through peer-reviewed publications, patents and presentations;
- improving the relevance of the knowledge position to certain business areas (units) measured through customer-satisfaction indices ('the business as customer');
- improving the position of existing products, in teams with marketing, measured by means of customer preferences;
- establishing networks and cooperation with outside specialists directly linked to specific business interest.

The aforementioned examples cover a large area but each can be made practical and can be formulated on normative, strategic and operational levels. It is important to try and avoid illusions of control, but always keep in mind that *progress measurement* will be made against goals.

Over the years numerous *tools* have been developed to facilitate the new product development process, to improve its effectiveness and to enhance its direction and focus. Nowadays hundreds of tools are available, representing experience from industry and consultancy as well as approaches from academic groups (Aranjo, 2001).

Based on a critical review of literature Den Ridder and Vernooy (2002) and Benner *et al.* (2003) define an NPD tool as a technique, methodology, procedure or model that is purposely constructed to improve the use, flow and quality of information as well as to assist the efficient execution of the NPD process. Following the classification of Gonzalez and Pulacios (2002), Benner *et al.* (2003) focused on the following groups of tools:
- tools for design such as quick product specification, quality function deployment, rapid prototyping, modular design and failure mode and effect analysis;
- tools for organisation such as concurrent activities management, stage-gate, multifactorial design teams;

- tools for manufacturing such as manufacturing resource planning, just in time, optimal products technology and statistical process control;
- tools for information technology such as computer-aided design, - manufacturing and engineering, computer-aided integrated manufacturing, internet and intranet, expert systems and product data management.

Tools are selected according to the specific needs of a business.

10.5.2. Portfolio management

For the implementation of the business strategy, portfolio evaluation has proven to be a powerful method for reviewing and prioritizing projects, to allocate resources to achieve corporate new product objectives and to select new science and technology areas and efficiently allocate resources to grow these areas. A key feature of portfolio evaluation is the presentation of complex information in a simple chart. On a *project* level this can be done by plotting the expected value of projects against the probability of success (Figure 10.6).

The evaluation of a project-portfolio is of most interest to those concerned with implementing the company's strategy in R&D. However, it is also important for those involved in projects to understand the context of their work. A number of conclusions can be drawn from the figures. Certain projects with high risk and lower expected value are less attractive than projects with low risk and higher expected value. The matrix enables management to consider resource reallocation by correlating the size of the project (the radius of the circle) to the position of the project within the matrix:

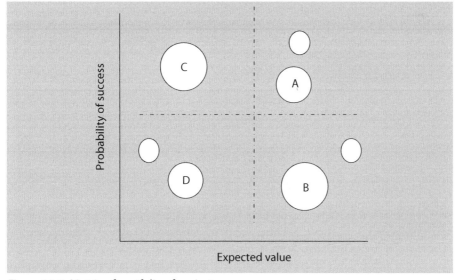

Figure 10.6. Matrix of portfolio of projects.

for instance, project A will be continued with an increased budget, project B will be continued but further focus on successful parts is recommended, project C will be continued with more focus and budget on value-creation and project D will be discontinued.

A key challenge to portfolio management is to ensure that the quality of the portfolio is such that it follows the lines of implementation of a company's strategy, that there are sufficient new ideas for the future, that an active management process is in place to create and encourage transitions to the segment of higher probability of success and higher expected value and that resources are adequately focused. Very often weak portfolio management translates into a huge reluctance to kill off new projects, poor evaluation methods and, in fact, the encouragement of weak projects.

There is a lack of rigorous decision points and a lack of stellar new product winners. On the whole, without portfolio management, the strategic criteria for project selection are missing (Cooper *et al.*, 2001). If applied on a consequential basis, portfolio management will lead to improved communication, a shared vision on strategic direction and shared ownership of a roadmap (10.5.3).

On the level of those involved in the *management of science and technology* it is important to ensure that a company's knowledge position is competitive and that the company is ready to meet future challenges. Following the principles of technology management (10.3.3) this can be achieved by regularly drawing a chart, shown in Figure 10.7, of the technological positions ('how good are we in certain technologies?') against the maturity of such technologies ('are we unique or is this technology accessible to everybody?').

This chart could lead us to relocate budgets ('the size of the circle') to new areas, to emphasize existing technologies in order to create more breakthroughs or to decide to exit certain areas. For instances, technology A could probably more effectively be explored in cooperation with outside partners, technologies B and C could benefit

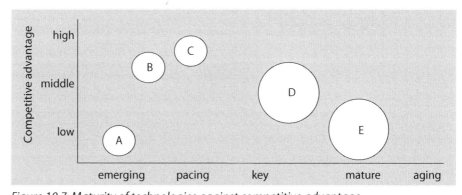

Figure 10.7. Maturity of technologies against competitive advantage.

from an increase in budget, technology D will continue at the same budget level, whereas the budget of technology E should be decreased.

Increasingly innovative solutions are being found for building platforms in emerging technologies, for instance in new, venture capital companies ('spin-outs') or in partnerships between existing companies.

An important question is which methods are used for measuring the rewards of research and development projects and investments. In order to assess economic rewards, criteria such as expected profitability (Net Present Value) or financial return (Internal Return Rate) are often used.

However, Cooper *et al.* (2001) concluded that strict reliance on financial methods for prioritizing projects is inappropriate: it is of greater importance to execute prioritisation decisions in the earlier phases of a project. It has been recommended that economic reward models are used as a piece of information in a larger process of determining value creation, which should certainly include the potential contribution to long-term strategy. The latter needs the involvement of a wider range of specialists (R&D, Marketing, Sales, Manufacturing, and Engineering).

10.5.3. Road mapping

A roadmap provides management's view on how to achieve a desired objective and is a framework for discussion between marketing, manufacturing, and science and technology (Figure 10.8). It leads to the integration of all aspects of technology into the business strategy. It integrates priorities from a product-portfolio exercise (10.5.2), technological reality and technological foresights. Road mapping facilitates a structured discussion of questions such as 'which technologies do we need in order to achieve a successful launch of a certain new product within one year? 'or 'which opportunities does a new development in technology bring to existing products?'.

The roadmap enables the translation of market goals into product strategies and provides clarity with regard to the number of products in the 'pipeline'. By tracing backwards the steps and intermediate goals it reveals the status and necessity of technologies and underlying research projects. This helps to prepare a five-year plan for financial expenses and human resources to meet corporate expectations on new products. By exploring possible future scenarios and identifying and quantifying risks and uncertainties, recognition will be obtained for opportunities and requirements for action. Road mapping is a continuous process and therefore is only worth the effort if it becomes part of a company's culture.

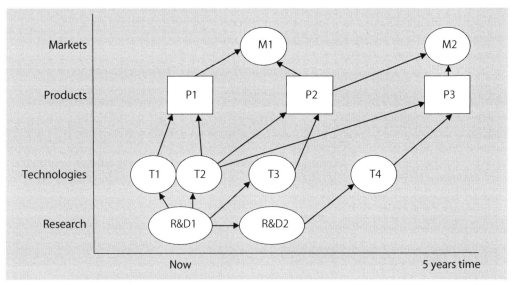

Figure 10.8. Roadmap of innovation process.

10.6. Concluding remarks

Like many other industries the food industry is constantly exploring opportunities for new products and added value. Modern techniques to optimise the role of knowledge in a business have become available. Although knowledge management is still in its infancy, basic principles and shared experiences enable us to establish some do's and don'ts for innovative companies. The objective of this chapter was to offer trends and general approaches that need further fine-tuning on the level of a particular business. In doing so, the following elements may present a map to find a way through the landscape of science, technology, business and innovation.

- Develop tools to assess strengths, weaknesses, opportunities and threats over the entire company as a basis for a business, product and project portfolio.
- Develop roadmaps of emerging technologies and business opportunities and a clear view of alternative futures: if needed, construct different scenarios and prepare contingency plans.
- Develop a strategy to achieve the vision, driven by the reality of product and technology platforms, partners, suppliers and customers and allocation of resources.
- Maximize the combined power of strengths.
- Manage the entire array of external technology and innovation sources.
- Align plans with a balanced portfolio of short-term and long-term interests.
- Ensure that the entire organisation is committed to creating and capturing value through innovation.

The nature of a company will determine the choices of strategy, the models used to manage the response to competitive and technological challenges, to business priorities and market trends. The food industry will need to balance its efforts between groundbreaking research and incremental developments and long-term and short-term needs. Important drivers are consumers' expectations, public attitudes, the regulatory environment and communication skills. A systematic approach in knowledge management as well as a commitment to science and technology and its inclusion in business targets will be the basis for an innovative food business.

References

Aranjo, J.S. (2001): Acquisition of product development tools in industry: a theoretical contribution. Technical University of Denmark, Lyngby.

Baeumner, A (2004). Nanosensors Identify Pathogens in Food. Food Technology 58 (8):51-56.

Benner, M. Geurts, R.F.R.., Linneman, A.R., Jongen, W.M.F., Folstar, P. and Cnossen H.J. (2003). A chain information model for structured knowledge management: towards effective and efficient food product improvement. Trends in Food Science & Technology 14:469-477.

Cook, S.D.N. and Brown, J.S. (2002). Bridging Epistemologies: The Generative Dance between Organizational knowledge and Organizational knowing. In: S. Little, P. Quintas and T. Ray (eds.) Managing knowledge Sage Publ. Londen.

Cooper, R.G., Edgett, S.J. and Kleinschmidt E.J. (2001). Managing Knowledge. Building Blocks for Success. Wiley, Chichester.

Cooper, R.G. and Kleinschmidt, E.J. (2000). New Product Performances: what distinguishes the star Products. Australian J. of Management 25 (1):17-46

De Rooy, J (2000). The consumer of the 21st century. In: Towards an Agenda for Agricultural Research in Europe. Wageningen Press, Wageningen, The Netherlands.

Den Ridder, C. and Vernooy, W. (2002). Insights into NPD-tools and their contribution to new product success. M.Sc.-Thesis Wageningen University.

Drejer, A. and Riis, J.O. (1999). Competence development and technology: how learning and technology can be meaningfully integrated. Technovation 19:631-649.

EIRMA (1999). The Management of Corporate Knowledge, European Industrial Research Management Association, Paris.

Folstar, P. (2001). Terug naar de basis. Naar een beter begrip van een veilige en duurzame voedselvoorziening. Wageningen University, The Netherlands.

Ganguly, A. (1999). Business-driven research and development. MacMillan Press. London.

Gonzalez, F.J.M and Palacios, T.M.B. (2002). The effect of new product development techniques on new product success in Spanish firms. Industrial Marketing Management 31:261-271.

Jonash, R.S. and Sommerlatte, T. (1999). The Innovation Premium, Perseus Books, Reading, Massachusetts.

Karel, M. (2000). Tasks of Food Technology in the 21st Century. Food Technology 54 (6):56-64.

Katz, R. (ed.) (1997). The Human Side of Managing Technological Innovation. Oxford University Press, Oxford.

Mark-Herbert, C (2004). Innovation of a new product category – functional foods. Technovation 24: 713-719.

Menrad, K. (2004). Innovations in the food industry in Germany. Research Policy 33:845-878.

Mermelstein, N.H. (2003). A look into the future of food science and technology. Food Technology 56 (1):46-55.

Montoya-Weiss, M.M. and Calantone, R. (1994). Determinants of new products performance: a review and meta-analysis. J. of Product Innovation Mnt. 11: 397-417.

Müller, M and Kersten S. (2003). Nutrigenomics: goals and strategies. Nature Reviews Genetics 4: 315-322.

Neumann, E and Thomas, J. (2002). Knowledge assembly for the life sciences. Drug Discovery Today 7 (20): S160-S162.

Newell, S., Robertson, M., Scarbrough, H. and Swan, J. (2002). Managing knowledge work. Palgrave, Basingstoke.

Probst, G., S. Raub and K. Romhardt (2002). Managing knowledge ; Building Block for Success. Wily & Sons, Chichester.

Roels, J. (2000). Food Processing: Challenges for Innovation. In: Towards an Agenda for Agricultural Research in Europe. Wageningen Press, Wageningen, The Netherlands.

Stewart-Knox, B. and P. Mitchell (2003). What separates the winners from the losers in new product development? Trends in Food Sciences & Technology 14: 58-64.

Tampoe, M. (1993). Motivating knowledge workers; the challenge for the 1990's. Long Range Planning 26(3) 49-55.

Tidd, J., Bessant, J. and Pavitt, K. (2001). Managing Innovation. Wiley, Chichester.

Trienekens, J.H. (2004). Quality and Safety in Food Supply Chains. In: Camps et al. (ed) : The Emerging World of Chains and Networks, Reed, The Hague.

11. Case study: development of a ready-to-eat meal for the health-conscious consumer

Marco Benner and Ruud Verkerk
Product Design and Quality Management Group, Wageningen University

11.1. Introduction

Over the past fifty years, the food industry has experienced many innovations in both processes and products. Technological developments in the fields of biotechnology, packaging, preservation, processing but also in the information technology, computer science and transport, are expected to change the future food business arena even more. It is vital for food companies to know whether a technologically new food product meets consumers' needs or will be accepted by the consumer in order to determine its economic feasibility. In this respect a major task lies in improving communication between scientists, technologists and consumers. To use the available knowledge in an efficient and effective way a structured approach to the food product development process is a necessity for food companies. To aid companies in realising such a structured approach the Chain Information Model has been developed (see Chapter 3). This model approaches the Food Product Development process from a chain-oriented point-of-view. The importance of chain cooperation for the food and agribusiness results from the specific characteristics of the food and agribusiness, e.g. (1) the limited shelf life of some products, (2) the natural variation in quality and quantity, (3) the variation in speed of the production process of the actors in the chain, (4) the differences in scale between the actors, (5) the complementary character of agricultural raw materials, (6) the intrinsic quality of fresh products, (7) the improved awareness of consumers towards food production systems, and (8) the need for and availability of capital (Zuurbier and Migchels, 1998; Den Ouden *et al.*, 1996).

The market for food products has changed drastically in the last few years. In most Western countries high production levels have led to saturated markets. Furthermore, demographic changes like an ageing population and changing household composition have had a major influence on the demand for food products. The role of the consumers has changed too. They are better educated and better informed than they used to be, resulting in higher requirements for product assortment and product quality. Moreover, the acceptance of a product no longer depends only on the quality of the product itself, but also on the way the product is produced. We can state in conclusion that food markets have shifted from a production orientation to a market orientation (Jongen *et al.*, 1996). In the last decade several developments have taken place in the food market, for example, the increased demand for healthy foods with a high level of convenience. The food industry has reacted to this development with the production of a variety of home meal replacements (HMR) varying from ready-to-cook to read-to-eat (Costa *et al.*, 2001). The share of these home meal replacements (HMR) in daily food consumption has increased enormously. The number of ready-to-eat meals and take away meals consumed in the Netherlands increased from 10 % in 1984 to 33 % in 2000 (Anonymous, 2000).

In this Chapter we demonstrate in a case study the development of a ready-to-eat meal containing vegetables (broccoli) with a certain health benefit.

11.2. Ready-to-eat meals

GfK Food Scan, a European market research organisation, differentiates between three groups of ready-to-eat meals: chilled, non-chilled (tinned, dried, etc.), and frozen. Within the group of chilled products a sub-division is made into day-fresh (prepared, packaged and cooled, with a maximum shelf life of 2 weeks), and chilled-fresh (prepared, packaged, pasteurised, and with a shelf life up to 6 weeks when stored refrigerated) (Zuurbier and Migchels, 1998).

The ready-to-eat meals market segment is growing every year. The individual markets in Europe show some differences, in volume as well as the type of meal consumed. Generally speaking the market for ready-to-eat meals is growing more in the north of Europe than in the southern part of Europe. In the Netherlands the growth of this segment, in volume, was 80% in 1996 compared to the growth in 1991. Of the 5 largest growing product groups in the supermarkets, the ready-to-eat meals segment came second. In Germany the growth of the ready-to-eat meal segment was as much as 124% (Zuurbier and Migchels, 1998). This impressive growth can be explained by several trends in society. Changes in life styles lead to changing eating habits. Both men and women are working and do not have the time, or do not want to use the time, to cook. This offers major opportunities to the producers of ready-to-eat meals (Zuurbier and Migchels, 1998).

Ready-to-eat markets have changed significantly since the first frozen TV-dinners were introduced; the application of new technologies has resulted in safer products with extended shelf life, better taste and higher nutritional value. And still consumers are not satisfied. They press for even safer and healthier food, it has to be more tasty, and they expect a larger variety of meals on offer. Moreover, the extrinsic product characteristics, as well as the way the ingredients are produced (animal friendly, no pesticides, no GMO's, etc.) are also becoming more and more important as regards consumer acceptance.

11.3. Health protective glucosinolates

The development of a ready-to-eat meal with a health benefit is based on the presence of a group of phytochemicals called glucosinolates in the vegetables that are a part of the meal. Glucosinolates are a group of secondary plant metabolites occurring in the *Cruciferae*, a family of plants that includes the Brassica vegetables such as cabbage, Brussels sprouts, broccoli and cauliflower. Glucosinolates co-exist with, but are physically separated from, the hydrolytic enzyme myrosinase in the intact Brassica plant. Upon mechanical injury of the tissue, the enzyme and substrate come into contact, resulting in hydrolysis. The features of the hydrolysis conditions such as pH, temperature and the presence of co-factors determine the proportion and nature of the various breakdown products. Substantial evidence suggests that these hydrolysed glucosinolate breakdown products possess important protective properties against cancer. This protective effect against cancer is caused by an induction of already existing protective detoxification systems in the human body. These protective compounds can reach the human body in two ways, namely directly by the consumption of breakdown products of glucosinolates hydrolysed by the myrosinase present in the plant, or indirectly by consumption of glucosinolates present in the vegetable and subsequently hydrolysed into protective compounds by the gut flora (Dekker *et al.*, 2000; Verkerk, 2002).

11.4. The production chain of ready-to-eat meals

Figure 11.1 depicts a simplified production chain for ready-to-eat meals. This chain consists of every actor that is physically in contact with the product or with components of the product. This does not include governmental bodies or consumer organisations, although they play a significant role in the acceptance of a product and hence in the product development process. Figure 11.1 shows that the production chain for ready-to-eat meals consists of many different actors, implying a complicated process of passing product and information. Breeders provide the basic input of the production chain, which may consist of seeds, young plants or young animals. Farmers produce the raw materials, they harvest the crop or breed the animals. Vegetables can be washed, cut or frozen. Potatoes can be peeled or processed into smaller parts. In this case processors are the producers of ready-to-eat meals. They check the raw materials, or semi-products, on arrival and prepare them for further processing. Heating (more or less severe), packaging and pasteurising are used for the processing. After pasteurization the package is chilled and prepared for order picking and transport. Distributors transport the chilled meals to the retailers where they are placed on the shelves. Finally, the product is sold and taken home by consumers, who heat the meal and consume it.

Each process in the production chain results in changes in the product and hence influences the (final) quality characteristics of the product. In order to control the final quality of the new or improved product the producer of ready-to-eat meals needs direct contact with all the different actors in the food production chain in order to negotiate what they have to produce (consumer-oriented product specification), what they can produce (production capacity) and how they have to produce it (what technologies they need).

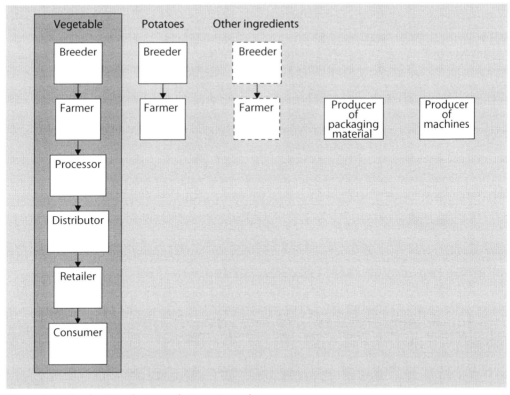

Figure 11.1. Production chain ready-to-eat meals.

In this case study we focus mainly on the technological information needed to produce the ready-to-eat meal with the optimal level of glucosinolates. Most of the information needed in the product development process has been taken from Dekker *et al.* (2000) on cabbage. They investigated the production circumstances that affect the glucosinolate content of cabbage. This study has shown that all actors in the production chain contribute to a considerable amount of variation in glucosinolate levels in Brassica vegetables at the moment of consumption. Suppose that a producer of ready-to-eat meals wants to innovate and aims at producing a health-protecting ready-to-eat meal. He chooses to make a broccoli-based meal, which he intends to promote for its high content of health-protecting glucosinolates. What are his options?

11.5. How to use the Chain Information Model?

To determine the options available for the realisation of the optimised product we have used the Chain Information Model, which exists of three phases: (1) the information-gathering phase, (2) the information-processing phase, and (3) the information-dissemination phase.

We approach the development process from the position of the meal producer, who acts as a chain-director and collects and distributes the information needed for successful development and production of the product.

11.5.1. Information-gathering phase

In the information-gathering phase, phase 1, the information needed for effective and efficient development of a new product is gathered. Three kinds of information are required to complete this phase. First, quality characteristics that make the current product successful are determined. These quality characteristics should not change in a negative way; therefore consumer demands regarding this product are assessed. Second, the current production chain, including actors and production processes, is mapped out. Third, information on the influences of the entire production chain on the new product feature is gathered. In practice, the desired quality characteristics should be determined by means of qualitative and quantitative consumer research. However, in this case study we used available data on consumer preferences from literature and earlier research conducted at Wageningen University and TNO Nutrition and Food Research in the Netherlands. In general, the main reason for consumers to buy ready-to-eat meals is convenience. Consumers either do not have the time for cooking their own meals, or they do not want to spend time on cooking (Datamonitor, 1998). Moreover, health is currently a popular topic for all food products. Consumers are more concerned than ever with their health and are looking for healthy foods (Euromonitor, 1996; Mintel, 1996). This health attribute is also sought for in the new generation of ready-to-eat meals that go beyond the TV dinners. Consumers are asking for the so-called chill-fresh meals that have undergone little or no industrial pre-heating. An often-criticised characteristic of ready-to-eat meals is the taste. According to the Central Agency for Food Products (CBL) in the Netherlands 60% of the consumers that buy ready-to-eat meals are not satisfied with the taste (Mintel, 1996). An additional consumer demand is the safety of the product (Euromonitor, 1996; Zuurbier and Migchels, 1998; Datamonitor, 1998; Samuelsson, 1999). Consequently, important primary consumer demands are convenience, healthiness, taste, and product safety. These consumer demands have to be made feasible for the product developers by translating them into operational, more specific demands (Table 11.1).

In order to find out what influences the actors have on the quality characteristics of the new product we have systematically determined what processes in the production chain had some kind of influence on each quality characteristic. We have used the technological expertise available within Wageningen University to determine these influences and placed them in so-called quality dependence diagrams to structure the process. Literature and expert knowledge within the university were used to determine the influence of the actors in the production chain on the New Product Feature (i.e. the glucosinolate concentration in the broccoli). This information is placed in a quality dependence diagram (Figure 11.2). In Figure 11.2 some actors are indicated with a dotted line, because these actors are supposed to follow the

Table 11.1. Consumer demands for ready-to-eat meals (Datamonitor, 1998; Euromonitor, 1996; Mintel, 1996; Samuelsson, 1999; Zuurbier and Migchels, 1998).

Convenience	short preparation time
	easy to open package
Healthiness	low fat content of the meal
	contains health promoting ingredients
	freshness of ingredients used
Sensory characteristics	mouth feel
	taste
	colour
Safety	microbial safety
	chemical safety
	physical safety

proper instructions coming with the product. The handling by these actors does not have a great influence on the product and the glucosinolate content will not be affected much. Moreover, the throughput time is rather high and therefore not much effect is expected. One exception to this is the household, since incorrect heating procedures can drastically lower the glucosinolate content. However, to a large extent the preparation cannot be controlled and influenced. We assume that this effect can be diminished by correct and clear instructions on the package.

With the information from the quality dependence diagrams, the information matrix can be completed (Figure 11.3). A distinction has been made between strong and weak relationships to indicate the difference between actors that can actively influence the glucosinolate content and actors that are supposed to follow instructions.

11.5.2. Information-processing phase

In phase 2 of the CIM, the information-processing phase, all the information collected in phase 1 is processed into essential information for the actors in the production chain. In this phase, there is an analysis of the influences of the processes in the entire chain, and hence the influences of the actors, on the new product feature and the quality characteristics of the product. The influences are placed in an information matrix that shows if, and to what extent, the actors influence the various quality characteristics.

Now that all the required information is gathered, this information has to be combined to determine the possible scenarios for the production chain to realise the intended product. These scenarios are formulated by a systematic analysis of the options every actor possesses to realise the required amount of glucosinolates in the meal. For every possible change made by an actor, the consequences for the

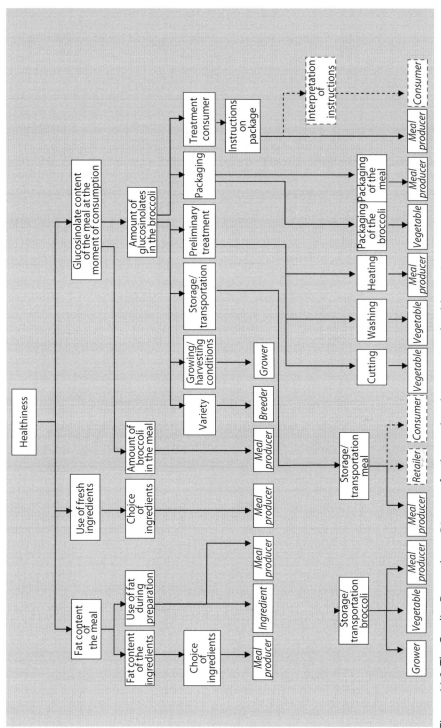

Figure 11.2. The Quality Dependence Diagram for the quality characteristic 'healthiness'.

Consumer demands	Actors	Breeder	Grower	Vegetable processor	Meal producer	Retailer	Consumer
Convenience	Preparation time	●		●	●		○
	Easy to open				●		
Healthiness	Fat content				●		
	Glucosinolate content	●	●	●	●	○	○
	Use of fresh ingredients				●		
Sensory characteristics	Taste (bitterness)	●	●	●	●		○
	Mouth feel	●	●	●	●		○
	Colour	●			●	○	○
Safety	Microbiological safety		●	●	●	○	○
	Chemical safety		●	●	●		
	Physical safety		●	●	●		

Figure 11.3. Information matrix.

other quality characteristics and the other actors have to be identified. We used decision trees to systematise the scenarios. To clarify the method we have elaborated the possibilities for two actors in the production chain, namely the breeder of the broccoli and the vegetable processor.

From Dekker *et al.* (2000) it is clear that the breeder has three options for raising the glucosinolate content: (1) by selecting an existing variety with a higher glucosinolate content, (2) by cross breeding, and (3) by genetic modification. By using a decision tree, each option can be checked for its usefulness (Figure 11.4). For each option the results for the other quality characteristics have to be examined before a choice can be made. First we suppose that the breeder has another broccoli variety at his disposal with the desired amount of glucosinolates. Before this variety can be applied in the ready-to-eat meal, the consequences for the other quality characteristics have to be checked. It is clear from Figure 11.3 that the breeder can influence the following quality characteristics: 'preparation time', 'taste (bitterness)', 'mouth feel', and 'colour'. For example, the preparation time of the meal can change because the new variety might have a more firm consistency resulting in a different mouth feel. Whether this change in preparation time is acceptable, has to be checked with the consumer, and therefore consumer acceptability intervals for the quality characteristics have to be defined. If there is no change in the preparation time or if the change is within the acceptability interval, the development team has to check the other quality characteristics for changes. If there is a change in the preparation time, the production chain has to be checked for solutions to counterbalance this change.

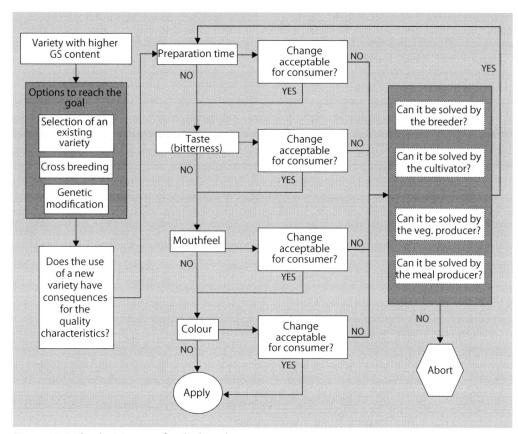

Figure 11.4. The decision tree for the breeder.

For instance, the cooking time of the broccoli will become shorter by cutting the broccoli in smaller pieces. This results in an information flow from the breeder to the vegetable processor. However, cutting the broccoli can also result in a loss of glucosinolates. Therefore, this solution has to be checked for its effects on the other quality characteristics to find a possible equilibrium. If the actors in the production chain cannot undo the changes, using this broccoli variety is not an option and other possibilities have to be considered.

The vegetable processor can influence the glucosinolate content of the broccoli by the procedures used for cutting, washing and storage (Dekker *et al.*, 2000). By cutting the vegetable, the enzyme myrosinase may come into contact with the glucosinolates and form breakdown products. Since these breakdown products are volatile they will be absent at the moment of consumption (De Vos and Blijleven, 1988). Washing conditions can influence the glucosinolate content in several ways, since the glucosinolates are soluble in water. For instance, the timing of the washing is relevant; whether the vegetable is washed before or after cutting can influence the loss. Temperature also influences the loss; hot water will result in a higher loss. The way in which the vegetable is washed can also influence the loss. In general, vegetables are washed by spraying or by dipping the vegetable into water. Dipping results in a longer contact time between vegetable and water and thus results in higher loss of glucosinolates (Verkerk, 2002). In addition, storage conditions influence the glucosinolate content (Hansen *et al.*, 1995). Before (one of) these options can be applied, the consequences for the consumer have to be examined. Figure 11.3 indicates that the vegetable processor can influence the consumer demands of 'preparation time', 'mouth feel', 'microbiological safety', 'chemical safety', and 'physical safety'. If the vegetable processor changes the cutting conditions by cutting the vegetable into larger pieces, this can influence the preparation time for the consumer. However, the meal producer can undo this by applying a mild heat treatment. Another way to counterbalance the longer preparation time is by applying a higher energy level during the microwave heating by the consumer, and therefore the information on the package has be adjusted to provide the consumer with the right preparation instructions. Figure 11.5 shows the systematic analysis for the scenarios for the vegetable processor.

This approach must be completed for all the actors that have a strong relationship with the quality characteristic 'glucosinolate content'. This results in several scenarios for raising the glucosinolate content in the production chain. In order to determine which scenario is the best, they have to be prioritised against criteria that are defined by the production chain. Important criteria for deciding what scenario is best can include financial feasibility and technological feasibility.

11.5.3. Information-dissemination phase

In the information-dissemination phase, phase 3 of the CIM, the information matrix from phase 2 is used to select the best scenario for creating the intended product. The selection is based on criteria which should be agreed on by the stakeholders. Once the best scenario has been selected it has to be implemented in the production chain, meaning that the proper information needed by each actor has to be exchanged.

11.6. Conclusion

The analysis of the case of glucosinolates in ready-to-eat meals shows that one needs to know where in the food production chain, how and to what extent the amount of glucosinolates can be influenced in order to develop a product with an added value: namely an optimised amount of health-protecting glucosinolates. Moreover, the case shows that food products are very complex systems with multiple interactions between ingredients and processes in the production process. This complexity generates an extensive flow of information that cannot be comprehended and managed by one company only. Therefore, the use of an inter-organizational tool, like the Chain Information Model, proves to be useful for the development of new

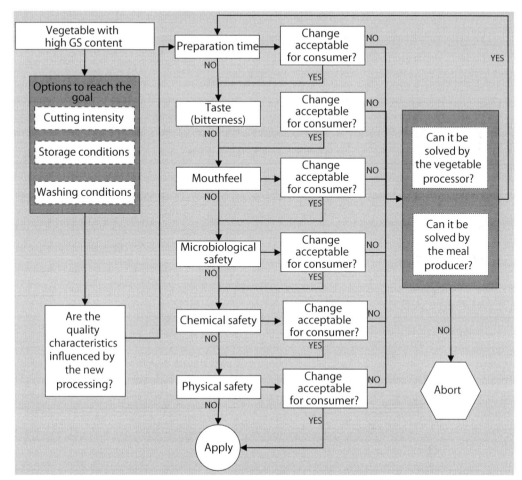

Figure 11.5. The decision tree for the vegetable processor.

products from a chain perspective. To make the information flow manageable it is recommended to use the chain-information model for gathering and structuring the information needed by each actor in the food production chain. In this way it is possible to create an effective, i.e. consumer-oriented, and efficient, thus structured and economical, product development process resulting in a ready-to-eat meal with health-protective properties as the New Product Feature.

All the information essential for the success of a new product has to be collected to use such a tool. Although the Chain Information Model mainly focuses on technological information other information is needed to decide which scenario is the best. For instance, financial information (what ingredients, packaging materials or processes should be used from a cost perspective), managerial information (strategic information, such as the company goals and the information a company is willing to share), governmental information (safety regulations, and labelling from a legislative perspective), and information from consumers and consumer organisations (what do consumers want, what ethical issues are at play, for instance, the use of GMO's).

The case study also shows that by a systematic analysis of changes in the production process, possible negative effects in terms of the quality characteristics that determine the success of the product can be predicted and possibly prevented. In this way, a higher success ratio can be achieved.

References

Anonymous (2000). Nederlanders eten steeds meer kant-en-klaar maaltijden. De Volkskrant, 18-10 2000.

Costa, A.I.A., M. Dekker, R.R. Beumer, F.M. Rombouts and W.M.F. Jongen (2001). A consumer-oriented classification system for home meal replacements. Food Quality and Preference 12: 229-242.

Datamonitor (1998). Ready meals: in-the-home convenience 1998. London: Datamonitor.

Dekker, M., Verkerk, R. and Jongen, W.M.F. (2000). Predictive modelling of health aspects in the food production chain: A case study on glucosinolates in cabbage. Trends in Food Science & Technology, 11 (4-5): 174-181.

Den Ouden, M., Dijkhuizen, A.A., Huirne, R.B.M. and Zuurbier, P.J.P. (1996). Vertical cooperation in Agricultural production-marketing chains, with special reference to product differentiation in pork. Agribusiness 12(3):277-290.

De Vos, R.H., and Blijleven, W.G.H. (1988). The effect of processing conditions on glucosinolates in cruciferour vegetables. Zeitschrift für Lebensmittel-Untersuchung und –Forschung (187): 525-529.

Euromonitor (1996). European markets: Chilled foods, delicatessen foods & ready meals, Euromonitor.

Hansen, M., Møller, P., Sørensen, H., and de Trejo, M.C. (1995). Glucosinolates in Broccoli stored inder controlled atmosphere. J. Amer. Soc. Hort. Sci., 120 (6): 1069-1074.

Jongen, W.M.F., Linnemann, A.R. and Dekker, M. (1996). Productkwaliteit uitgangspunt bij aansturen product(ie)technologie vanuit keten (in Dutch). Voedingsmiddelentechnologie 26: 11-15.

Mintel (1996). Chilled ready meals. London: Mintel International Group Limited.

Samuelsson, M. (1999). The Dutch market for chilled ready meals. Zeist: TNO.

Verkerk, R. (2002). Glucosinolate levels throughout the production chain of Brassica levels. Towards a novel predictive modeling approach. PhD-thesis, Wageningen University, Wageningen, 136 pp.

Zuurbier, P.J.P. and Migchels, N.G. (1998). Nutri 2000: Haalbaarheidsstudie naar de productie van kant-en-klaar maaltijden in Noord Nederland (in Dutch). Kluwer: Wageningen.

Epilogue

By the end of this book, the reader has learned about the many technological facets of product and process design. Food product and process *development* is a collaborative effort of people working together in a team towards a common goal: the production of a food that complies with consumer wishes. The *design* phase is a critical aspect of product development, but there is of course more to it than that. As mentioned in the introduction, knowledge of consumer behaviour and marketing is indispensable, as are cost constraints and legislation. The reader is advised that the process of food product development is certainly not a linear one; there are frequent iterations and control loops that sometimes make the outcome of the process difficult to predict. The results of the technological design phase must be frequently checked with general management, marketers, economists, with people knowledgeable in legislation, and with people responsible for environmental aspects. As a result of these interactions, the design will probably have to be optimized. This can only be done efficiently if the whole process is supported wholeheartedly by general management, if there is open communication and trust, in other words, if the team is helped to do what it has to do: design a product that fulfils a consumer need. It is hoped that this book has contributed to this process by highlighting the technology part of the design phase.

Keyword index

A

acidification 237
action-oriented culture 246, 250
activation energy 84, 85, 93, 94, 96
active packaging 201, 202
activity coefficient 110, 111
actors in the production chain 59, 60, 64, 265, 267, 268, 271
advertising 243
Akaike criterion 82
allocation 237
analogy 36, 41, 42
antimicrobial
 – agent 129, 138
 – emitters 201
antioxidant emitters 201
antioxidants 138
Antoine equation 109, 122
Arrhenius
 – equation 84, 85, 86, 94, 95, 98
 – law 83
aseptic packaging 219
association 36, 38, 40, 41, 42

B

bacterial growth 69
bacteria-tight 217, 220, 221
Bancroft rule 158
barrier
 – formation 138
 – properties 128, 198
Bayesian belief networks 67, 103
BET isotherm 112
Bigelow model 74, 87, 92, 93
bi-layer system 125
binary mixture 113
biofilms 217, 218
biopolymer 127, 129
bio-surfactants 128
biphasic behaviour 89
biphasic inactivation 90
blow-up 179

brainstorming 28, 33, 34, 35, 36, 39, 43, 48, 49
 – unguided 34
brand development 243
break-up mechanism 170
bridging 160
business creativity 34
butterfly thinkers 39, 42

C

capillary number 174, 177
caramelization 69
casein 159
ceramic membrane 171, 186
CFD *See:* computational fluid dynamics
chain
 – approach 19
 – information model (CIM) 59, 60, 61, 263, 266, 272, 273
 – reversal 17, 67
change perspective 42
CIM *See:* chain information model
coacervation 131
coalescence 151, 153, 154, 157, 158, 159, 160
coating 107, 108, 120, 123, 125, 126, 130, 133, 134, 135, 136, 140
commitment trust 254
companion trust 254
company knowledge 28
competence trust 254
competitive assessment 56
computational fluid dynamics (CFD) 103, 180, 189
computer-aided design 257
concept development 33, 39, 43
conceptual model 73
conducive conditions 30
consumer
 – orientated approach 19
 – perception 123, 125, 137
 – types 21

consumption
 – situation 21
 – time 21
contamination 208, 211, 212, 213,
 214, 219, 220
continuous phase 152, 156, 158, 159,
 162, 163, 166, 168, 171, 172, 173,
 174, 176, 177, 179, 181, 184, 188
convenience 197, 202, 267
convenience consumer 24
correlation matrix 55
corrosion 216, 217
creaming 154, 156, 159, 162, 163
creation 30
creative thinking techniques 32, 33,
 36, 37, 39, 41
cross contamination 208, 211, 215
cross-flow 171, 172, 173, 174, 175,
 176, 177, 187
crystallization kinetics 90
curing 127
curvature 154, 178, 179
customer competitive assessment 56

D
d3,2 *See:* sauter diameter
deamidation 69
decision
 – making process 99
 – support systems 103
 – support tool 229, 230
 – tree 62, 63, 64, 270, 271, 273
decontamination 216, 218, 219, 220
demographic changes 17, 263
denaturation 69, 89, 96
depletion flocculation 160, 161
deterministic model 74
diffusion 70, 71, 75, 85, 94, 112
 – coefficient 198
diffusivity 114, 115, 118, 119, 120,
 121, 122, 123, 128, 137
dip casting 132
dispersed phase 152, 154, 158, 162,
 163, 168, 171, 172, 177, 178, 179,
 180, 181, 182, 183, 190

DLVO theory 160
droplet 152, 158, 159, 164, 166, 167,
 168, 172, 173, 174, 176, 177, 178,
 179, 180, 182, 187, 188, 189
 – size 151, 154, 156, 162, 163, 170,
 172, 174, 177, 179, 180, 181, 182,
 185, 186, 188, 189
 – size distribution 155
duplex emulsion 153

E
eco-indicators 238
eco-label 239
E. coli 218
EDGE *See:* Edge-based Droplet
 GEneration
Edge-based Droplet GEneration
 (EDGE) 176, 181, 182, 185, 186
edible
 – coating 107, 137
 – films 108
Eilers equation 90
Einstein 89
emissions 228, 235, 237
empirical model 73
emulsifier 152
emulsion 124, 128, 151, 152, 153, 156,
 159, 164
 – stability 156, 159, 162, 163
energy 151, 153, 154, 156, 168, 169,
 170, 185, 186
 – density 165, 175
engineering competitive assessment
 56
environmental impact 228
environment-conscious consumer 24
environment policy 228, 229
enzymatic
 – browning 68, 87
 – polymerization 69
enzyme
 – activity 69, 70
 – catalysed reactions 96
 – inactivation 89
 – kinetics 87, 88

– barrier   128, 140
– migration   128, 137, 139
– regulators   201
mould growth   69
multiresponse modelling   81
mycotoxins   209

N
nature and animal-loving consumer
   24
neural networks   67, 103
new preservation technologies   20
new product development (NPD)   29,
   256
Newtonian behaviour   130
non-enzymatic browning   68, 69, 70,
   77
nonlinear regression   79
non-thermal preservation techniques
   21
normalization   238
NPD   *See:* new product development

O
objective target values (HOW
   MUCHs)   56
ohmic heating   21
Ostwald ripening   127, 163
oxidation   67, 69, 70, 97, 200
oxygen scavenger   201

P
packaging   107, 238
parallel model   121
parameter
– D   86
– Q10   86
– Z   86
pasteurization   215, 217, 220
patents   242
permeability   115, 116, 118, 120, 121,
   122, 123, 124, 125, 126, 127, 128,
   129, 138, 139, 199
permeance   115, 133, 139
permeation   134

phospholipid   152
photochemical reactions   97
physical interaction   69
pickering emulsions   151, 159
plasticizer   129, 134, 139
Platonic and Socratic influences   37
polyethylene glycol   129, 134
polymer   156, 158, 160, 162, 187, 188
portfolio evaluation   257
precision of parameters   83
predictive microbiology   74
predictive model   74, 203
pre-exponential factor   84, 85, 93, 96
pre-mix emulsification   172, 173, 174,
   175, 188, 190
price-conscious consumer   24
probabilistic models   98
probiotics   28
Product Planning Matrix   54, 58
product requirements (HOWs)   55
profitability   64
protein   152, 158
proteolysis   68
pulsed electric fields   21

Q
QDD   *See:* quality dependence
   diagrams
QFD   *See:* quality function
   deployment
quality dependence diagrams (QDD)
   62, 63, 267
quality function deployment (QFD)
   53

R
radical reactions   97
radio frequency identification (RFID)
   203
Raoult's law   70
rate constants   85
rational techniques   41
reaction kinetics   86
recontamination   208, 209, 215, 219,
   221

relationship matrix 55, 57
resource
 – depletion 237
 – oriented methodologies 48
retrogradation 69
RFID *See:* radio frequency
  identification
rheological properties 90
risk
 – analysis 220
 – assessment 214, 255
road mapping 259
rotor-stator system 164

S
safety 20, 74, 138, 197, 198, 201, 202,
  204, 207, 216, 223, 244, 267
sauter diameter 156
scalability 187
scenarios 59, 60, 62, 63, 272
secondary coating 126
second generation brainstorming 48
second-order
 – model 82, 83
 – reaction 80, 86
sedimentation 159, 160, 162, 163
shared imagination 33
shelf life 18, 20, 72, 74, 91, 94, 98, 99,
  107, 108, 138, 139, 197, 200, 201,
  214, 215, 219, 221, 263, 264
Shirazu Porous Glass 171
simulation 178
SIT *See:* systematic inventive thinking
socially-conscious consumer 24
societal setting 17
solvent-based casting 131
solvent evaporation 131
sorption isotherm 112, 114, 136, 139
span 156
SPG membrane 186
spontaneous droplet formation 184,
  185
spores 212, 216
square root model 97
stabilisation 197

stability map 161
stabilizer 151, 153, 155, 156, 158
Stage-Gate approach 255
sterilization 215, 216, 217, 220
stochastic model 74, 204
Stokes Law 162
Strengths and Weaknesses,
  Opportunities and Threats
  (SWOT) 35, 40
successful products 31
surface area 154, 156, 183
surfactant 136, 151, 156, 157, 158,
  174, 188
susceptor materials 202
sustainability 228
swan thinkers 39
swarm of droplets 163
SWOT *See:* Strengths and
  Weaknesses, Opportunities and
  Threats
systematic experimental design 102
systematic inventive thinking (SIT)
  48

T
tacit knowledge 245
technical importance ratings 57
technological feasibility 64
technology platforms 246, 251, 260
templates 49
 – attribute dependency 49
 – division 49
 – multiplication 49
 – subtraction 49
 – task unification 49
thermobacteriology 86
thermodynamics 108, 109, 120, 138
time-temperature indicator (TTI)
  203, 204
T-junction 176, 177, 190
total quality management (TQM) 53
toxins 209
TQM *See:* total quality management
trust 253, 254
TTI *See:* time-temperature indicator